Heisenberg and the Interpretation of Quantum Mechanics

Werner Heisenberg was a pivotal figure in the development of quantum mechanics in the 1920s, and also one of its most insightful interpreters. Together with Bohr, Heisenberg forged what is commonly known as the 'Copenhagen interpretation'. Yet Heisenberg's philosophical viewpoint did not remain fixed over time, and his interpretation of quantum mechanics differed in several crucial respects from Bohr's. This book traces the development of Heisenberg's philosophy of quantum mechanics, beginning with his positivism of the mid-1920s, through his neo-Kantian reading of Bohr in the 1930s, and culminating with his 'linguistic turn' in the 1940s and 1950s. It focuses on the nature of this transformation in Heisenberg's thought and its wider philosophical context, which have up until now not received the attention they deserve. This new perspective on Heisenberg's interpretation of quantum mechanics will interest researchers and graduate students in the history and philosophy of twentieth-century physics.

KRISTIAN CAMILLERI is a Lecturer in the History and Philosophy of Science Program at the University of Melbourne, Australia.

T0296524

Werner Heisenberg (*Werner-Heisenberg-Archiv*, Munich). Reproduced with the permission of H. Rechenberg.

Heisenberg and the Interpretation of Quantum Mechanics

The Physicist as Philosopher

KRISTIAN CAMILLERI
University of Melbourne

CAMBRIDGE
UNIVERSITY PRESS

CAMBRIDGE UNIVERSITY PRESS
Cambridge, New York, Melbourne, Madrid, Cape Town,
Singapore, São Paulo, Delhi, Tokyo, Mexico City

Cambridge University Press
The Edinburgh Building, Cambridge CB2 8RU, UK

Published in the United States of America by Cambridge University Press, New York

www.cambridge.org
Information on this title: www.cambridge.org/9781107403512

First published 2009
First paperback edition 2011

A catalogue record for this publication is available from the British Library

Library of Congress Cataloguing in Publication data
Camilleri, Kristian.
Werner Heisenberg and the interpretation of quantum mechanics / Kristian Camilleri.
p.   cm.
Includes bibliographical references and index.
1. Heisenberg, Werner, 1901–1976 – Philosophy.   2. Physics – Philosophy.   3. Quantum
theory.   1. Title.
QC16.H395.C34   2008
530.12–dc22
2008034113

ISBN 978-0-521-88484-6 Hardback
ISBN 978-1-107-40351-2 Paperback

To my parents, Rita and Joseph

Contents

Preface   *page* ix

1   Introduction   1
1.1   Heisenberg's philosophy of quantum mechanics   3
1.2   Heisenberg as a philosopher-physicist   8

PART I   THE EMERGENCE OF QUANTUM MECHANICS   15

2   Quantum mechanics and the principle of observability   17
2.1   The observability principle   18
2.2   The renunciation of the electron orbit and the elimination of
      unobservables   21
2.3   Heisenberg's version of the observability principle   26
2.4   The meaning of observability: a discussion with Einstein   31
3   The problem of interpretation   36
3.1   Reconceptualising the electron's motion   37
3.2   The physical meaning of Schrödinger's wave mechanics   40
3.3   Drawing the battle lines: Heisenberg and Schrödinger on
      understanding in physics   44
3.4   Redefining *Anschaulichkeit*   48
3.5   Instrumentalism or realism? Heisenberg's notion of closed
      theories   53

PART II   THE HEISENBERG–BOHR DIALOGUE   61

4   The wave–particle duality   63
4.1   The physical reality of the electron: wave or particle?   65

4.2	Probability waves and the quantum mechanics of particles	68
4.3	Quantised matter waves and wave–particle equivalence	73
4.4	Heisenberg and Bohr: divergent views of wave–particle duality	77
5	**Indeterminacy and the limits of classical concepts: the turning point in Heisenberg's thought**	**85**
5.1	New concepts of space and time in quantum theory	87
5.2	The introduction of the gamma-ray microscope	92
5.3	Bohr's analysis of the limits of measurement and the meaning of concepts	95
5.4	The turning point in Heisenberg's philosophy	100
6	**Heisenberg and Bohr: divergent viewpoints of complementarity**	**108**
6.1	Bohr's complementarity of space-time and causal descriptions	109
6.2	The Heisenberg interpretation	112
6.3	Weizsäcker's reconstruction	116
6.4	Mutually exclusive experimental arrangements	118
6.5	The object–instrument divide	123
	PART III HEISENBERG'S EPISTEMOLOGY AND ONTOLOGY OF QUANTUM MECHANICS	131
7	**The transformation of Kantian philosophy**	**133**
7.1	Heisenberg's early confrontation with Kantian philosophy	134
7.2	The doctrine of classical concepts	139
7.3	The constitutive dimension of language and the forms of intuition	142
7.4	Heisenberg's transformation of the a priori	147
8	**The linguistic turn in Heisenberg's thought**	**152**
8.1	Heisenberg's conception of language	153
8.2	The language-reality problem in quantum mechanics	158
8.3	The problem of meaning	161
8.4	Actuality and potentiality	165
	Conclusion	**172**
	References	179
	Index	196

Preface

When I began work on my doctoral dissertation my aim was not to embark on a historical study of Heisenberg's philosophical thought, but rather to investigate the concept of indeterminacy in quantum mechanics. My intention initially was to begin the thesis with an introductory chapter focusing on the writings of Bohr and Heisenberg, whose discussions in Copenhagen in 1926–7 formed the basis for what would later become known as the 'Copenhagen Interpretation' of quantum mechanics.

However, as I became more closely acquainted with Heisenberg's interpretation of quantum mechanics, I became more intrigued by his philosophical viewpoint. I quickly arrived at the conclusion that what little had been written on the subject had not really done justice to the depth or originality of Heisenberg's interpretation of quantum mechanics. Indeed, much of the secondary literature tended to dismiss him as a positivist, or portray him in his later years as adopting some form of neo-Platonism. What I found particularly striking about the existing secondary literature was the absence of any discussion of Heisenberg's relationship to Kant's philosophy and his emphasis on the constitutive dimension of language, in spite of the fact that both of these are recurring themes in his later writings. A more detailed study of Heisenberg's philosophy of quantum mechanics therefore appeared to me as valuable from both a historical and a philosophical perspective.

It was with this in mind that I subsequently decided to make Heisenberg's philosophical viewpoint the central focus of my thesis. In 2005, I completed my doctoral dissertation entitled 'Heisenberg and Quantum Mechanics: The Evolution of a Philosophy of Nature'. Encouraged by one of my examiners, Professor Don Howard, and my colleagues at the University of Melbourne, I subsequently undertook the task of turning the dissertation into a book. This work is the product of that effort. The ideas contained in Heisenberg's seminal paper on matrix mechanics in 1925 mark the beginning not only of a new phase in modern physics, but also of Heisenberg's own philosophical journey. This book examines that journey as it unfolded between 1925 and the late 1950s by situating Heisenberg's philosophy of quantum mechanics within the context of various strands of thought in the German-speaking world at the time.

Heisenberg's early inclination towards positivism in the 1920s bears the decisive influence of Einstein's theory of relativity. Yet, by the late 1920s, Heisenberg's thought had moved away from the empiricist viewpoint that had underpinned his early philosophy of quantum mechanics. The nature and scope of this transformation, which forms the central theme of this book, has, up until now, been poorly understood and often completely neglected. Instrumental in this regard were Heisenberg's discussions with Bohr in Copenhagen in 1927. The recognition of the indispensability of classical concepts forms the point of departure for much of Heisenberg's later thought, which brought him into contact with the attempts in the German-speaking world of the 1930s to unravel the basic problem of Kantian epistemology. Increasingly, Heisenberg saw the problem of reality in quantum mechanics under the aegis of a philosophy of language, according to which concepts of classical physics are accorded a priori status insofar as they are the conditions for the possibility of objective experience in spite of their limited applicability. The concept of an object existing in space and time is for Heisenberg an idealisation, but one which remains indispensable for the objectivity of knowledge. By the 1950s, Heisenberg argued that quantum mechanics describes a world of possibilities and potentialities, which paradoxically cannot be regarded as 'objectively real' in any ordinary sense of the term.

While Bohr exerted perhaps the most important philosophical influence on Heisenberg, their intellectual relationship was characterised by hidden disagreements and misunderstandings. This is most strikingly displayed in their respective views on wave–particle duality and complementarity. While Heisenberg certainly used terms such as 'complementarity' and 'wave–particle duality' in his writings, a close reading reveals that these terms had very different meanings for the two physicists. In bringing to light these divergences between Bohr and Heisenberg, this book lends further weight to the view that what is commonly referred to as the Copenhagen interpretation of quantum mechanics comprises a number of different viewpoints and philosophical positions.

A book of this scope and kind is not possible without the support, assistance and encouragement of many people. My primary thanks go to my doctoral supervisor, Dr Keith Hutchison, whose comments and criticism were invaluable in writing the original thesis. I am also indebted to Professor Don Howard, with whom I had many stimulating conversations about the work during my time as Visiting Fellow at the Center for Science, Technology and Values at the University of Notre Dame in 2007. I owe a debt of gratitude to Dr Vera Butler and Dr Gerhard Wiesenfeldt for their assistance in translating several difficult passages from German into English. While acknowledging their assistance, I take full responsibility for any inaccuracies of translation that appear in this book. Thanks must also go to Edward Hare, who helped with references and citations in the final stages of the preparation of the manuscript.

On a personal note, I would also like to extend my thanks to Giovanna Bartolo for her kind hospitality during May of 2004, in allowing me to work virtually uninterrupted on the thesis, during my stay at her property in Elphinstone. I would also like to thank Howard Sankey, Kelly Farrell, and my parents Joseph and Rita Camilleri for their continual encouragement and helpful comments after proofreading through drafts of the chapters. Without them I have no doubt this book would never have been written. I would also like to acknowledge the contribution made by my friends Costandinos Khtorides, Dr Stephen Ames and Sergio Mariscal, whose intellectual stimulation and encouragement I greatly appreciated during the time of writing the thesis. I would also like to extend my thanks to the anonymous reviewers of the proposal I submitted to Cambridge University Press for their constructive comments as well as the referees of the papers I submitted for publication based on chapters of my original dissertation. Their comments and criticisms have been crucial in revising and rewriting certain sections of the book.

The material from the Hans Reichenbach Collection contained in the *Archives of Scientific Philosophy* has been quoted with the permission of the University of Pittsburgh. I am also grateful to the Special Collections staff at the Ballieu Library at the University of Melbourne for granting me access and permission to quote extensively from the *Archive for the History of Quantum Physics*, which has proved to be an invaluable resource for this work.

A substantial portion of this book derives from papers that have already appeared, although most of the chapters were revised before becoming part of the book. I would therefore like to thank Elsevier, Taylor and Francis and MIT Press for granting me permission to reproduce the material which appears in those chapters. Here is the list of the papers.

- Chapter 4 is based on 'Heisenberg and the Wave–Particle Duality', *Studies in the History and Philosophy of Modern Physics*, 37 (2006), pp. 298–315. Reprinted by permission of Elsevier.
- Chapter 5 is a slightly modified version of 'Indeterminacy and the Limits of Classical Concepts: The Turning Point in Heisenberg's Thought'. *Perspectives on Science*, 15 (2) (2007), pp. 176–99. Reprinted by permission of MIT Press.

- Chapter 6 is an extended and revised version of 'Heisenberg, Bohr and the Divergent Viewpoints of Complementarity'. _Studies in the History and Philosophy of Modern Physics_, 38 (3) (2007), pp. 514–28. Reprinted by permission of Elsevier.
- Chapter 7 is based on 'Heisenberg and the Transformation of Kantian Philosophy', _International Studies in the Philosophy of Science_, 19(3) (2005), pp. 271–87. Reprinted by permission of Taylor and Francis.

I have employed the following abbreviations in the references throughout the book:

ASP = Archive for Scientific Philosophy
AHQP = Archive for the History of Quantum Physics
BSC = Bohr's Scientific Correspondence

1

Introduction

The philosophical problems posed by quantum mechanics were evident to many of the founders of the theory from its inception in the mid-1920s. More than 80 years later, many of these issues continue to generate much debate and disagreement. Yet we are only just beginning to understand the history of the interpretations of quantum mechanics, through the careful study of the work of leading physicists who figured prominently in the classic debates of the 1930s. According to the standard view found in much of the secondary literature, the discussions between Niels Bohr and Werner Heisenberg in Copenhagen in the first half of 1927 laid the foundations for what is commonly referred to today as the 'Copenhagen interpretation of quantum mechanics'. At the 1927 Como and Solvay conferences Bohr presented his concept of complementarity, which in his view provided a new conceptual framework for understanding quantum mechanics. This view, which Bohr would develop further in the late 1920s and 1930s, is frequently understood to be the philosophical foundation of the Copenhagen interpretation of quantum mechanics, which called for a dramatic revision of the hitherto accepted foundations of physics and epistemology. As Dugald Murdoch explains, 'By the end of 1927 the Copenhagen interpretation had established itself as the dominant interpretation of quantum mechanics' (Murdoch, 1994, p. 303). While there were dissenting voices, notably those of Einstein, Schrödinger and Planck in the 1930s, it is widely recognised that Bohr's views prevailed in the decades that followed and became the basis of the 'orthodox view'.

But precisely what the Copenhagen interpretation is, or what it tells us about the world of atoms and electrons, turns out to be a rather difficult question to answer. As Susanne Gieser points out, 'Many attempts have been made to characterize and analyse the philosophical and epistemological position of the Copenhagen School and especially of Bohr, and its significance in the emergence of the definitive interpretation of quantum mechanics' (Gieser, 2005, p. 56). Indeed, despite an extensive literature which discusses and criticises the so-called

'Copenhagen interpretation', and the philosophical position which underpins it, there remains no agreement on what it is. The lack of clarity about precisely what constitutes the Copenhagen interpretation stems partly from the fact that none of the founding fathers of quantum mechanics ever set out in a clear fashion the basic tenets of the orthodox interpretation. Indeed, while Bohr's concept of complementarity was hailed by many of his contemporaries as 'the most significant result for philosophy that crystallized out of modern physics' (Jordan, 1944, p. 131; see also Pauli, 1980, p. 7; 1994), Bohr's writings were interpreted through a variety of different philosophical perspectives ranging from logical positivism (Jordan, 1936, p. vii; Frank, 1957, pp. 216–17; 1975, pp. 162–5) and neo-Kantianism (Weizsäcker, 1971a, 1994) to dialectical materialism (Rosenfeld, 1979a) and subjective idealism (Blokhintzev, 1952). Moreover, von Neumann, Dirac and Wigner – all of whom made significant contributions to the development of the orthodox interpretation of quantum mechanics – either completely ignored or were explicitly critical of Bohr's notion of complementarity (Bub, 1995). As Max Jammer points out in his *Philosophy of Quantum Mechanics*,

> the Copenhagen interpretation is not a single, clear-cut, unambiguously defined set of ideas but rather a common denominator for a variety of related viewpoints. Nor is it necessarily linked with a specific philosophical or ideological position. It can be, and has been, professed by adherents to most diverging philosophical views, ranging from strict subjectivism and pure idealism through neo-Kantianism, critical realism, to positivism and dialectical materialism.
>
> *(Jammer, 1974, p. 87)*

In a similar vein, Erhard Scheibe warns us 'that there is no point in looking for *the* Copenhagen interpretation as a unified and consistent logical structure' (Scheibe, 1973, p. 9). The term 'Copenhagen interpretation', Scheibe argues, refers to the divergent, and sometimes conflicting, views of physicists 'who played an important role in the establishment of quantum mechanics, and who were collaborators of Bohr's at his Institute or took part in discussions during the crucial years'. This view has more recently found support in the works of John Hendry (1984, p. 1), Catherine Chevalley (1999, pp. 173, 189), Mara Beller (1999, pp. 9, 173) and Don Howard (2004). As each of these authors suggests, beneath the veneer of agreement one finds hidden discord and debate between the various adherents of the so-called 'Copenhagen interpretation'. In order to arrive at a deeper understanding of the history of interpretations of quantum mechanics we must look beyond the supposed 'unity of the Copenhagen school' and focus on the views of the individual physicists.

This book contributes to just this task, by examining the philosophical interpretation of quantum mechanics of one of the most important physicists of the Copenhagen school – Werner Heisenberg (1901–76). Born in Würzburg, Heisenberg studied the new quantum theory in Munich under Arnold Sommerfeld in the early 1920s. In the winter of 1922–3 he spent time in Göttingen as Max Born's assistant before gaining his *venia legendi* at Göttingen University. He also spent several months at the Institute for Theoretical Physics in Copenhagen in 1924–5 and again in 1926–7, where he collaborated closely with Bohr. During this period, Heisenberg produced two of his most important works – his seminal paper on matrix mechanics, which he published in July 1925, and his celebrated paper on the uncertainty principle, published in March 1927. Heisenberg was among the brightest young stars in a new generation of theoretical physicists, who were instrumental in the creation of quantum mechanics in the 1920s. In recognition of his contribution, he was awarded the Nobel Prize in physics in 1932. By the age of 26 Heisenberg had secured a full professorial appointment as a lecturer in Theoretical Physics at Leipzig University. He would go on to become one of the pioneers of quantum electrodynamics and quantum field theory in the 1930s, and make important contributions to the field of nuclear physics before turning his attention to the search for a unified field theory of elementary particles in the 1950s (Hermann, 1977; Kleint & Wiemers, 1993). Heisenberg's impact on theoretical physics in the twentieth century was the subject of two symposia in 2001 in commemoration of the centenary of his birth (Papenfuβ, Lüst & Schleich, 2002; Buschhorn & Wess, 2004).

Heisenberg is remembered today as one of the principal architects of quantum mechanics, but he was also one of its most insightful interpreters. He brought an unusually profound grasp of the philosophical problems involved, matched only in depth and significance by Bohr. Heisenberg devoted considerable attention to the philosophical foundations of quantum mechanics and wrote extensively on that subject. However, his writings on the epistemological and ontological questions, which underpinned his interpretation of quantum mechanics, have received only scant attention and remain the source of considerable misunderstanding and ambiguity. This work attempts to address this lacuna in Heisenberg scholarship and the history of the interpretations of quantum mechanics.

1.1 Heisenberg's philosophy of quantum mechanics

To speak of Heisenberg's philosophy of quantum mechanics is somewhat misleading. Heisenberg's philosophical viewpoint did not remain fixed throughout his lifetime, but underwent significant transformation in the late

1920s and 1930s. In focusing on the way in which Heisenberg's thought evolved over time, this book brings to light a number of the key themes that emerged in his writings between 1925 and 1960. Here the historical development of Heisenberg's thought can be properly understood only by situating it in the context of his discussions with other physicists such as Bohr and Einstein, as well as his contact with various philosophical schools of thought in the German-speaking world, all of which left their mark on his thought. While I think that a deeper appreciation of Heisenberg's thought offers something of value to those interested in the philosophy of quantum mechanics, the task of the present work should be understood primarily as historical, rather than as a contribution to contemporary debates in the philosophy of physics. Indeed, Heisenberg left a number of key issues unresolved or ambiguous in his later writings. To this extent, my aim in writing this book has not been to defend Heisenberg's philosophical position, but merely to understand it.

In attempting to piece together a coherent picture of the development of Heisenberg's philosophy of quantum mechanics, this work draws on two different historiographical approaches. The first is the 'dialogical historiography' outlined by Mara Beller and John Hendry, which seeks to emphasise 'the complex dialogical nature of thought' in the history of science (Beller, 1999, p. 3). In his biography of Heisenberg, David Cassidy suggests that he depended on conversations with 'his friends and colleagues for philosophical stimulus' (Cassidy, 1992, p. 48). Taking up this approach, one finds that his philosophy of quantum mechanics was shaped not by private meditation on the problems posed by the new theory, but rather by his dialogues, encounters and correspondence with other physicists and philosophers of his time, primarily with Bohr but also with Einstein, Schrödinger, Pauli and Weizsäcker. The second approach is taken from the work of Catherine Chevalley, who has argued that we cannot fully understand the original interpretations of quantum mechanics without appreciating the precise philosophical context in which they developed. She declares that '*history of science* and *history of philosophy* are equally necessary' (Chevalley, 1994, p. 50, emphasis in original). In taking this perspective I examine the historical development of Heisenberg's thought by situating it against the background of the divergent reactions to Kant's philosophy in the German-speaking world in the 1920s and early 1930s. Only through an integration of the history of science and the history of philosophy, mediated by discussions with his contemporaries, can we more adequately grasp and appreciate the manner in which Heisenberg's philosophy of quantum mechanics took shape over time.

This work is divided into three parts. Each of these corresponds roughly to a historical phase in the development of Heisenberg's thought, though there is considerable overlap between them. Part I of this work deals with Heisenberg's philosophical viewpoint which arose during the *emergence* of quantum mechanics in the period between 1925 and 1927. In particular, this section is devoted to shedding light on two principal themes explained in Chapters 2 and 3: the observability principle, which is widely thought to have inspired his ground-breaking 1925 paper on quantum mechanics; and the subsequent clash between Heisenberg and Schrödinger in 1926–7 over what it means to *understand* a physical theory, which underpinned the different attitudes to the problem of interpretation in matrix mechanics and wave mechanics.

Part II of the book deals with a number of themes which emerge in the context of *Heisenberg's dialogue with Bohr* in the period between 1926 and 1930 concerning the interpretation of quantum mechanics. Chapters 4, 5 and 6 examine three key ideas that arose in discussions with Bohr during this critical period: (i) the notion of the wave–particle duality; (ii) the concept of indeterminacy and the limited applicability of classical concepts; and (iii) Bohr's viewpoint of complementarity. In each of these chapters I have attempted to trace not only the ways in which Heisenberg's thinking on these issues owed a debt to Bohr, but also how his point of view marks a departure from Bohr's. Part III focuses on the final phase in the development of Heisenberg's *epistemology* and *ontology* of quantum mechanics. By the mid-1930s Heisenberg had begun to view the problem of reality in quantum mechanics as inextricably linked to a reformulation of Kant's notion of the *a priori* and the world-disclosing function of human language.

What emerges from this study of the historical development of Heisenberg's philosophy is a complex but decisive shift from a broadly empiricist outlook to a philosophical viewpoint which embraces the constitutive dimension of human language. The earlier phase of Heisenberg's thought bears the influence of Einstein's theory of relativity, or, to be more precise, the strongly positivistic tendency which many of Heisenberg's contemporaries, Einstein included, considered integral to the theory. This is evident in three ways: first, in his introduction of the principle of observability into quantum mechanics in 1925; secondly, in an instrumentalist view of understanding in physics and the subsequent redefinition of *Anschaulichkeit* (visualisability); and thirdly, in an operational analysis of kinematic concepts in 1927. In each of these cases, Heisenberg attempted to draw an explicit or implicit analogy with the philosophical lessons of the theory of relativity, or, to put it more precisely, the positivistic attitude which he felt had underpinned the Einsteinian conception of space and time. However, as I argue in Chapter 3, Heisenberg's instrumentalism requires more careful attention. Although he described the task of physics to uncover empirically adequate mathematical laws, he also took the view that once a theory had reached the status of a 'closed theory' in physics – through an axiomatic structure which described a wide range of empirical phenomena – as quantum mechanics had by 1927, one could interpret the theory as in some way representing the form or structure of reality itself and not merely as a phenomenological description. In this sense Heisenberg's philosophical view was in some way closer to 'structural realism' than to instrumentalism.

We can see the beginnings of the shift away from positivism in Heisenberg's thought in his discussions with Einstein and Bohr in 1926–7. Here Heisenberg recognised the problematic nature of the concept of observability (discussed in Chapter 2) and the operational point of view (discussed in Chapter 5). The abandonment of operationalism, which had played an important role in his analysis of the gamma-ray microscope thought experiment in the 1927 paper on the uncertainty relations, in particular, marked an important turning point in Heisenberg's thought. After discussions with Bohr, Heisenberg arrived at the view that it was not possible, as he had previously thought, to replace classical concepts like position and momentum with new quantum concepts in the description of experimental phenomena. By the 1930s, influenced by discussions with his student and friend Carl Friedrich von Weizsäcker and the visiting Kantian scholar Grete Herman, Heisenberg began to stress the importance of Kant's philosophy for understanding quantum mechanics. Yet Heisenberg was no orthodox Kantian. As I show in Chapter 7, his later epistemology is characterised by a transformation of Kant's notion of the *a priori*. The classical forms of intuition of space and time are for us the conditions for the possibility of all experience, but at the same time they have only limited applicability. Moreover, such forms do not, for Heisenberg, originate in 'pure reason' but emerge historically through our interplay with the world. In this sense, space and time prove to be indispensable as conditions for the possibility of empirical science, and yet they are not transcendental in Kant's strict sense.

By the 1940s the positivism characteristic of his early approach had been abandoned in favour of a different epistemological view in which the world-disclosing function of language assumes central importance. Here we find that Heisenberg situates the paradoxes of quantum mechanics within what I have termed a 'quasi-transcendental', as opposed to an analytic, conception of language. As Heisenberg would put it, somewhat paradoxically, 'for us there is only the world in which the expression "there is" has meaning'. The beginnings of this philosophical attitude can be traced back to his critical exchanges with Bohr in Copenhagen in 1927, which centred on the indispensability of classical concepts. Here Heisenberg came to the realisation that while quantum mechanics demanded an abandonment of the classical analytic concept of motion, certain forms of classical thought such as space and time were indispensable for a description of experience, even in quantum mechanics. Rather than replacing the concepts of classical physics with new quantum concepts, or with operational definitions, Heisenberg now resigned himself to the fact that we cannot dispense with the concepts of classical physics, in spite of their limitations. Here Heisenberg would depart from his earlier emphasis on the elimination of unobservables and an operational definition of concepts, recognising that 'we are suspended in language'. By the 1950s one can discern in Heisenberg's philosophy his own 'linguistic turn' according to which 'objective reality' has meaning for us only within the framework of space and time. Somewhat paradoxically then, the quantum world cannot be deemed 'objectively real', but is merely a world of 'possibilities' or 'potentialities'.

Notwithstanding Bohr's immense influence on Heisenberg, the latter's epistemology remained distinct from those of his Danish colleague on several critical points. While scholars such as Beller and Howard have emphasised the point that Bohr and Heisenberg disagreed in important ways, little attention has been devoted to disentangling their respective views of wave–particle duality and complementarity (Beller, 1999, p. 9). This is precisely the task of Chapters 4 and 6. Heisenberg's understanding of the wave–particle duality is based on the formal *symmetry* or *equivalence* of wave and particle descriptions, not the necessity of using them in mutually exclusive experimental arrangements in Bohr's sense. Indeed, while Heisenberg often presented himself as an enthusiastic proponent of Bohr's concept complementarity, his view of the complementarity of space-time and causal description is based on a misunderstanding of a crucial passage in Bohr's Como lecture. Indeed, whereas Bohr emphasised that mutually exclusive experimental arrangements serve to define the conditions for the *unambiguous* use of classical concepts such as position and momentum, Heisenberg was inclined to see this situation as highlighting the inherent *ambiguity* in the use of classical concepts in the complementarity description. Drawing on a careful textual analysis of Bohr's and Heisenberg's writings, as well as recent scholarship which has clarified certain aspects of Bohr's notion of complementarity, we can see more clearly the hidden, but nonetheless, substantial differences in their respective philosophical positions on the interpretation of quantum mechanics and the defence of the 'completeness' of quantum mechanics in the 1930s. Such divergences have gone largely unnoticed, as they were often buried within Heisenberg's carefully worded exposition of these ideas.

It is important, however, not to overstate the divergences between Heisenberg and Bohr. Mara Beller, in her important book *Quantum Dialogue*, argues that Heisenberg only ever paid lip service to Bohr's doctrine of the indispensability of classical concepts, but that he did not subscribe to this view consistently

(Beller, 1999, pp. 182, 197–9). However, in my view, Beller's conclusion is based on a misunderstanding of the crucial passages in which Heisenberg articulated his position. The accusations of inconsistency frequently levelled at Heisenberg, while sometimes warranted, are in this case based on a failure to subject his writings to careful scrutiny. While on several occasions Heisenberg drew a contrast between the ambiguity of our everyday language and the precision of the mathematical description in physics, after 1930 he never deviated from the view that when asked to describe the *results* of our measurements 'we are forced to use the language of classical physics, simply because we have no other language in which to express the results' (Heisenberg, 1971, pp. 129–30). A close reading of the relevant texts shows that by the late 1920s Heisenberg became convinced that classical concepts were indispensable for a description of experience in quantum mechanics, though his reasons for holding this view, and the epistemological viewpoint which eventually underpinned it, differed from Bohr's.

1.2   Heisenberg as a philosopher-physicist

In the words of Don Howard, 'The fifty years from 1880 to 1930 was the era of the philosopher-physicist' (Howard, in press). Here Howard identifies Einstein as the leading figure in 'a whole generation of scientists' including Bohr, Heisenberg, Pauli, Schrödinger and Weyl, all of whom could be 'equally well described as philosopher-physicists'. Yet the image of Bohr as the philosophical leader of the Copenhagen school has meant that Heisenberg has remained largely in Bohr's shadow. In 1965, Patrick Heelan declared, 'the epistemology of quantum mechanics has up to now been studied almost exclusively through the works of Bohr', whereas 'Heisenberg's philosophy has been curiously untouched' (Heelan, 1965, pp. ix–x). Heelan's *Quantum Mechanics and Objectivity: A Study of the Physical Philosophy of Werner Heisenberg*, which was published in 1965, remains to my knowledge the only major study published in English of Heisenberg's philosophy of physics. By contrast, the last 30 years have witnessed a continuation of scholarly interest in Bohr's philosophy of quantum mechanics in the English-speaking world, much of which has attempted to locate Bohr's views in the framework of the realism debate in the philosophy of science (Folse, 1985; Honner, 1987; Murdoch, 1987; Faye, 1991; Favrholdt, 1992; Faye & Folse, 1994).

As one might expect, Heisenberg's philosophy has attracted more attention from scholars in continental Europe than in the Anglophone world, as is evident in Herbert Hörz's *Werner Heisenberg und die Philosophie* (1968) and Guiseppe

Gembillo's *Werner Heisenberg: la filosofia di un fisico* (1987). More recently the proceedings of two conferences held in 1991 and 2001 devoted to Heisenberg's physics and philosophy have been published (Geyer, Herwig & Rechenberg, 1993; Gembillo & Altavilla, 2002). While these works offer some valuable insights into Heisenberg's philosophy of physics, they remain virtually unknown in the English-speaking world. There has, however, been a new surge of interest in Heisenberg's philosophy of physics over the last decade or so, particularly concerning his notion of the 'closed theories' in physics. The recent work of scholars like Alisa Bokulich (2006) and Melanie Frappier (2004) has done much to re-examine some of the key ideas in Heisenberg's philosophical thought. Beyond this the works of Catherine Chevalley (1988), Carl Friedrich von Weizsäcker (1971b, 1987) and Jan Lacki (2002), though by no means constituting systematic studies of Heisenberg's interpretation of quantum mechanics, provide useful insights into some of the central themes in Heisenberg's philosophy of physics.

It is important to note that this book differs from the major studies undertaken by Heelan, Hörz and Gembillo, both in approach and in scope. Heelan frames his reading of Heisenberg largely in terms of Husserl's transcendental phenomenology. Though he quotes extensively from Heisenberg's writings, Heelan's primary aim is to give his own philosophical interpretation of quantum mechanics, rather than engaging in serious historical scholarship. In attempting to trace the development of Heisenberg's thought, Heelan argues that after 'an early and predominantly empiricist phase', Heisenberg 'passed to a predominantly rationalist viewpoint ... inspired almost totally by the transcendental philosophy of Kant' (Heelan, 1965, pp. xiii–xiv). While Heelan quite rightly recognises that Heisenberg moved away from an early empiricist phase, he fails to appreciate the different forms this empiricism assumed in Heisenberg's early work, and his subsequent critique of the observability principle and the operational standpoint in the late 1920s. Moreover, his claim that Heisenberg's later thought is best described as a 'rationalist viewpoint' is somewhat problematic. Neo-Kantian philosophy certainly exerted an important influence on Heisenberg, but his later writings are explicitly critical of the rationalist viewpoint espoused by Kant, and embrace a far more pragmatic outlook largely influenced by Bohr and Weizsäcker. Indeed, in a letter to Pauli in 1935, Heisenberg commented that he found the dissertation by the neo-Kantian scholar Grete Hermann to be 'reasonable', though perhaps too much inclined to 'the rationalist philosophical tendency' (Heisenberg to Pauli, 2 July 1935, Pauli, 1985, p. 408 [item 414]). Heisenberg saw quantum mechanics as having brought about a pragmatic revision of Kant's notion of *a priori* knowledge – a theme which I elaborate on in Chapter 7.

While useful in drawing attention to the otherwise neglected aspects of Heisenberg's thought, the books of Hörz and Gembillo pay little attention to the historical development of Heisenberg's interpretation of quantum mechanics, preferring to focus on his relationship with the various philosophical traditions with which he came into contact. Furthermore, these works do not limit themselves to an investigation of the philosophy of quantum mechanics but examine Heisenberg's later philosophy on the unified field theory of elementary particles, which increasingly drew him towards some kind of neo-Platonism (Hörz, 1968, pp. 220–68; Sallee, 1983; Gembillo, 1987, pp. 1–65). By contrast, this work does not address the search for a unified field theory, which constituted the central task of Heisenberg's physics after 1950, but confines itself solely to Heisenberg's philosophical understanding of quantum mechanics. To this extent this book can only be but a first step in exploring Heisenberg's overall worldview. A comprehensive account of Heisenberg's philosophical vision would require a more detailed examination of his later neo-Platonism, his enigmatic notion of the 'central order', as well as his views on the relationship between science, culture and religion.[1] A comparison of the scientific and religious worldviews of Heisenberg and Planck is to be found in the recent work of Cornelia Liesenfeld (1992) and Wilifred Schröder (1999).

Despite Heisenberg's extensive writings on epistemological questions, there has been a tendency to portray Heisenberg as having contributed little of significance or originality to the philosophy of quantum mechanics. In his biography, David Cassidy argues that as a physicist Heisenberg exercised only 'a modest, usual interest' in philosophy (Cassidy, 1992, p. 255). This view finds some support from Weizsäcker, who portrays Heisenberg as someone who saw himself first and foremost as a physicist, not a philosopher, and to this extent remained somewhat reluctant to immerse himself in the epistemological problems of modern physics (Weizsäcker, 1985, p. 184). When Heisenberg did turn his attention to philosophy, 'his interest ... was primarily neither ontological nor epistemological but one of an aesthetic nature' (Weizsäcker, 1987, p. 287). Yet at the same time Weizsäcker acknowledges that 'Heisenberg's was a philosophical mind'. Indeed, Heisenberg had hoped to co-author a book with Weizsäcker on the 'philosophical relevance of modern physics', through an examination of different schools of thought such as materialism, positivism, Thomism, critical idealism, Hegelianism and Platonism (Weizsäcker, 1971b, p. 11). Though the book was never written, Heisenberg's published writings give the distinct impression of a thinker preoccupied with the philosophical implications of

---

[1] Gregor Schiemann from the University of Wuppertal has recently completed a book on Heisenberg's philosophy which focuses on some of these themes (Schiemann, 2008).

modern physics. Indeed, three volumes of his collected works are devoted to his epistemological writings (Heisenberg, 1984c, 1984d, 1985a). It is worth noting that many of Heisenberg's philosophical writings are in the form of lectures presented to audiences without formal training in physics or philosophy. Thus he rarely, if ever, presented his work in the language or style of academic philosophy. His *Wandlungen in der Grundlagen der Naturwissenschaft*, which was a collection of his more philosophical lectures of the 1930s and 1940s, went through numerous editions and was translated into many languages (Heisenberg, 1949). Probably the best known of Heisenberg's popular philosophical works, *Physics and Philosophy*, is a publication of his Gifford lectures, presented at St Andrews, Scotland, during the winter semester of 1955–6.

Though Heisenberg never aspired to reach the lofty heights of academic philosophy, he possessed an impressive knowledge of the history of philosophy, and his grasp of the western philosophical tradition far exceeded that of Bohr. A careful survey of Heisenberg's writings reveals some familiarity with the writings of Plato, Aristotle, William of Ockham, Duns Scotus, Thomas Aquinas, Descartes, Kant, Fichte, Hegel, Mach and Wittgenstein.[2] However, there is little evidence to suggest that Heisenberg ever studied any of these authors in a systematic fashion. Perhaps more important for Heisenberg's philosophical thought was his contact with many of the leading intellectuals of his day.[3] In the 1930s he engaged in discussions and correspondence with philosophers as different as Moritz Schlick and Martin Heidegger on the epistemological and ontological lessons of quantum mechanics. During this period Heisenberg was brought into contact with a number of the leading figures of logical positivist movement, who were interested in the philosophical implications of the new theory. Yet we also know that in autumn 1935, he was Heidegger's guest in Todtnauberg, where he had participated in philosophical discussions lasting several days (Weizsäcker, 1988, p. 230). Heisenberg continued to correspond with Heidegger over the next 40 years, and in 1959, contributed an article to the Heidegger *Festschrift* (Heisenberg, 1959; Hempel, 1990). Patrick Heelan, whose interpretation of Heisenberg situates him within the phenomenological tradition, goes so far as to suggest that 'the modern European continental philosopher feels closer to him in spirit than does,

---

[2] I have not included Goethe in this list, though it is clear that Heisenberg saw Goethe as a deeply philosophical thinker, whose worldview was one of the sources of inspiration for his own view of reality (Partenheimer, 1989, pp. 55–77). Goethe's influence on Heisenberg is clearly evident in the philosophical manuscript of 1942.

[3] As Catherine Carson has argued, by the 1930s Heisenberg had succeeded Planck as one of the heirs of the *Kulturträger* in West German intellectual life, a role that brought him into contact with many prominent German thinkers (Carson, 1995).

perhaps, his Anglo-American counterpart' (Heelan, 1965, p. ix). While Heelan's reading of Heisenberg is undoubtedly coloured by his own philosophical orientation, there is certainly a sense in which Heisenberg became increasingly hostile to the analytical tradition in philosophy. On several occasions in his later years he declared his preference for the later Wittgenstein over the Wittgenstein of the *Tractatus* and expressed his disagreement with Bertrand Russell on the philosophy of language (Stapp, 1972, p. 1115; Peat & Buckley, 1996, pp. 7–8). In a letter written to Hans Reichenbach reporting on the 1936 'Unity of Science' congress in Copenhagen, Martin Strauss observed that

> Heisenberg, who was here at the conference, has generally speaking, a strange attitude to philosophy. He abuses the positivists, something that is not without justification, but he does so from "right". Good philosophy = a great personality who has something to say, like Heidegger (who nevertheless is not great enough).
> *(July 1936, Hans Reichenbach Collection [HR-013-35-07] ASP. Quoted by permission of the University of Pittsburgh. All rights reserved)*

Intriguing though this comment is, I can find nothing to support Catherine Chevalley's suggestion that 'Heisenberg found in Heidegger's interpretation of modern ontology the most powerful expression of his own views about the signification of quantum mechanics' (Chevalley, 1988, pp. 172–3). While Heisenberg remained in sporadic correspondence with Heidegger until the 1970s, there is no evidence to indicate that the latter's ontological standpoint exerted any influence on his interpretation of quantum mechanics. Indeed, Heisenberg was on occasions quite critical of what little he understood of Heidegger's philosophy. In a letter, written in 1973, to Heidegger congratulating him on his eightieth birthday, Heisenberg took issue with Heidegger's view of the hermeneutic-ontological relationship between the human being and the world (Heidegger, 1977, pp. 44–5). The points of convergence between Heisenberg's outlook and twentieth-century German philosophy are not to be found in any systematic adherence to the phenomenology of Husserl or Heidegger. Rather, they are to be located in a general attitude to the task of philosophy, and more specifically in Heisenberg's later conception of the constitutive dimension of human language. Here, one sees a striking similarity to the 'linguistic turn' in German philosophy, which originated with Hamann, Herder and Humboldt in the early nineteenth century, and was subsequently taken up in different ways by Heisenberg's contemporaries, perhaps most notably in the work of Ernst Cassirer.

The clearest expression of Heisenberg's underlying philosophical impulse in this later period is to be found in his untitled private philosophical manuscript written around 1942, which was published posthumously in 1984 in his collected works as 'Ordnung der Wirklichkeit' (Heisenberg, 1984e, pp. 217–306). The 1942 manuscript, as I will refer to it, was deemed by Heisenberg to be 'too personal' to be published in his own lifetime.[4] Yet it serves as an extremely important text, shedding new light on some of the more enigmatic passages in Heisenberg's later writings. The manuscript was translated into Italian in 1991 by Giuseppe Gembillo, and, more recently, Catherine Chevalley has published a French translation, with a detailed commentary and introduction (Heisenberg, 1991, 1998). In the light of its significance, it is worth noting that the manuscript is not cited in the works of Heelan and Hörz, both of which were published in the 1960s well before it first appeared in print. Gembillo's 1987 study also fails to refer to the manuscript. Its importance for a proper understanding of Heisenberg's thought is emphasised by Chevalley, who describes it 'the densest and most synthetic elaboration of Heisenberg's ideas on epistemological significance of contemporary physics and on the problem of knowledge in general' (Chevalley, 1998, p. 11).

The manuscript makes clear the extent to which in his later years Heisenberg had departed from the positivism characteristic of his earlier thinking, and adopted an epistemological viewpoint centring around the constitutive dimension of language. Though the text is important in making sense of much of Heisenberg's later thought, it is important to note that it is not a work of epistemology. Indeed, as Catherine Chevalley rightly points out, the manuscript is as much a consolatory reflection on the social, political and cultural situation in which Heisenberg found himself in Nazi Germany during the Second World War, as it is an insight into Heisenberg's philosophical worldview (Chevalley, 1998, pp. 11–13). To this extent, an exhaustive study of this fascinating text would take us far beyond the scope of this work, which as I stated earlier remains confined to his philosophy of quantum mechanics.

In endeavouring to make sense of Heisenberg's philosophical thought, much is to be gained from paying careful attention to Heisenberg's correspondence with other physicists and philosophers as well as to his interviews, particularly in his later years. While Heisenberg's later recollections of past conversations and events should be treated with an element of suspicion, his interviews with Kuhn and others in the 1960s and 1970s serve to clarify a number of important points which are expressed somewhat ambiguously in his published writings. Through a careful examination of the way that Heisenberg's philosophy unfolded in the period from the 1920s to the 1950s, it becomes clear that despite the immense influence exerted by Bohr, Heisenberg's philosophical viewpoint

---

[4] On 10 February 1947 Heisenberg wrote to Fritz Klauss, requesting that the manuscript not be published because it was 'too personal' (Chevalley, 1998, p. 10).

stands as an original and unique contribution to the interpretation of quantum mechanics, worthy of careful study. A deeper appreciation of his view should be of interest not only to historians and philosophers of quantum mechanics, but also to scholars working on European intellectual history of the twentieth century. Heisenberg's more philosophical views on quantum mechanics reflected not only his understanding of the theory and his ongoing dialogue with other physicists on its interpretation, but also his receptivity to many of the important contemporary currents in philosophy in the German-speaking world.

PART I

The emergence of quantum mechanics

2

Quantum mechanics and the principle of observability

Heisenberg's July 1925 paper '*Über quantentheoretische Umdeutung kinema-tischer und mechanischer Bezeihungen*' (On the Quantum-Theoretical Reinterpretation of Kinematic and Mechanical Relations) marks the birth of modern quantum mechanics. It is also well known for introducing into quantum theory an 'observability principle', which asserts that only quantities that can in principle be observed should be included in physics. This idea had gained currency in the 1920s among many of Heisenberg's contemporaries, notably Pauli, Born and Jordan, largely because of its perceived importance in Einstein's theory of relativity in abolishing the classical notions of absolute space and time. To this extent, the philosophical viewpoint from which quantum mechanics emerged is generally understood to be that of positivism. As the Austrian physicist Hans Thirring would put it in 1926: 'Quantum mechanics; the theory founded by Heisenberg, and then further extended by Born, Jordan and Dirac, can be considered as a kind of phenomenological theory, as it poses for itself the task of establishing relations only between quantities that are observable in principle' (Thiring, 1928, p. 385). Though Heisenberg's early papers on quantum mechanics suggest that he embraced a positivistic attitude, a closer examination of the relevant texts gives a more nuanced view of the role of positivism in Heisenberg's early thought. In the first instance, the observability principle did *not*, as is commonly thought, form the driving motivation behind Heisenberg's elimination of the concept of the electron orbit in quantum theory. Rather, it appears that Heisenberg included the principle in the introduction to his paper, only after he had derived the formal stricture of the theory in an effort to give some kind of philosophical legitimacy to his radical new approach. To this extent, the observability principle appears in the history of quantum mechanics, not as a guiding principle, but as a *post facto* justification for the elimination of the electron orbit.

This situation forces a re-evaluation of the role of positivism in Heisenberg's early thinking. Undoubtedly, in appealing to the 'principle of observability',

Heisenberg attempted to give his work a distinctively 'positivist' character, which would hopefully make his elimination of classical particle trajectories more agreeable to his contemporaries. Indeed, Pauli had earlier complained that Heisenberg's approach to physics was 'unphilosophical'. It therefore seems entirely reasonable to suggest that the introduction of the observability principle by Heisenberg in his 1925 paper, while not playing a central role in the development of quantum mechanics, did reflect his efforts to 'legitimise' his approach to quantum theory. However, if we focus on the way in which Heisenberg understood the observability principle a still more complicated picture emerges. Here I want to argue that Heisenberg's introduction of the observability principle into quantum theory actually brought to the fore the *problematic* nature of the concept of observability. This point has escaped the attention of many authors who have read the introduction to Heisenberg's paper in which we find his first explicit statement of the observability principle, as mere repetition of the earlier views of Pauli, Born and Jordan. A careful reading of the introduction to the paper, and the subsequent references to observability in the mid-1920s, however, suggests otherwise.

As we shall see, Heisenberg had originally argued that we are entitled to conclude that 'hitherto unobservable' quantities such as the electron's position and period of its orbit are unobservable *in principle*, because their inclusion into quantum theory has failed to realise a consistent and empirically adequate theory. However, after his discussion with Einstein in April 1926, Heisenberg became aware that even this pragmatic version of the 'observability principle' was susceptible to critique. Einstein was able to convince Heisenberg that only once we are in possession of a complete theory can we determine what is 'observable'. In the absence of such a 'complete' theory of quantum mechanics, Heisenberg conceded that it was not yet possible to state with any confidence whether the electron's position was 'unobservable in principle' or not. This shift marks an important retreat from the positivistic inclinations of the 1925 paper, to a more critical position with regard to the concept of observation in physics. In effect, the introduction of the observability principle into his 1925 paper on quantum mechanics, though giving it a decidedly positivist appearance, ultimately led Heisenberg to call into question the sharp distinction between theory and observation, and in so doing, marked the beginning of his turn *away* from the prevalent positivistic attitude that characterised much of the scientific philosophy of the time.

2.1 The observability principle

It is first necessary to examine the philosophical background to the introduction of observability into quantum mechanics. In his seminal paper on quantum mechanics written in July 1925, Heisenberg declared that his aim was 'to establish a basis for theoretical quantum mechanics, founded exclusively upon relationships between quantities which in principle are observable' (Heisenberg, 1925a, p. 879; 1967a, p. 261).[1] The introduction of the principle of observability into quantum theory has thus understandably been the subject of much discussion among historians and philosophers of science. As Mehra and Rechenberg put it, 'the idea is often considered as the philosophical basis of Heisenberg's theory and Heisenberg is held responsible for having introduced it as a guiding principle into quantum theory' (Mehra & Rechenberg, 1982b, p. 274). This view is echoed in Heelan's 1966 study of Heisenberg's philosophy of quantum mechanics: 'The great insight which brought about the discovery of quantum mechanics was that physics should concern itself only with *observable quantities*' (Heelan, 1965, p. 29). These accounts indicate that the observability principle played a pivotal role in Heisenberg's abandonment of the classical picture of the orbiting electron, which was crucial to his breakthrough into quantum mechanics.

As many historians of science have rightly noted, the principle of observability was the subject of much discussion around the time that Heisenberg published his 1925 paper. Pauli, Born and Jordan, all of whom were in close contact with Heisenberg at the time, were prominent advocates of the idea that unobservable quantities should be eliminated from physics. This idea was traced back to Mach and found its most powerful expression in Einstein's theory of relativity. By the 1920s, it was widely accepted that Einstein's critique of the concepts of absolute space and time had been greatly influenced by his reading of Mach's work. Indeed, in 1916, Einstein had explicitly acknowledged his intellectual debt to the empiricist-positivist tradition, in particular to Hume and Mach (Einstein, 1916). Ernst Cassirer makes the point abundantly clear in his 1921 philosophical study of the theory of relativity: 'It is precisely this principle of "observability", which Einstein applied at an important and decisive place in his theory ... Any physical explanation of a phenomenon, he urges, is *epistemologically* satisfactory only when there enter into it no non-observable elements' (Cassirer, 1923, p. 377, emphasis in original). In the 1920s, physicists often stressed the significance of Einstein's great insight, not least Wolfgang Pauli, whose classic work on the theory of relativity written in 1921, stated the principle with remarkable clarity: 'we should adhere to the idea that in physics only quantities that are in principle observable should be introduced' (Pauli, 1958, p. 206). Pauli's standpoint was to a large extent inspired by Einstein's analysis of the concepts of space and time in the special theory of relativity, and

---

[1] I will refer to the English translation of Heisenberg's 1925 paper (Heisenberg, 1967a) which appears in B.L. van der Waerden except where otherwise indicated.

would exert quite a considerable influence in Göttingen, where Heisenberg, Born and Jordan developed matrix mechanics in the second half of 1925.

In a paper published in 1919, Pauli had criticised Hermann Weyl's recent attempts to develop a unified field theory, because it made use of quantities that were unobservable in principle. Weyl had supposed that an electron is nothing other than a region of space in which energy density of the matter field is extremely high, and that it was to this extent possible to define the strength of the electric field even *inside* an electron. For Pauli, on the other hand, the concept of the electric field strength in the interior of the electron was a *meaningless* concept, since there was no conceivable measurement that could be performed to determine the magnitude of the field strength at a point inside the electron. As Pauli put it in his paper:

> In Weyl's theory we continuously operate with the field strength in the interior of the electron. For a physicist this is only defined as a force on a test-body, and since there are no smaller test-bodies than the electron itself, the concept of electric field strength in a mathematical point [in the region inside the electron] seems to be an empty meaningless fiction. One should stick to introducing only those quantities in physics that are observable in principle.
>
> *(Pauli, 1919, pp. 749–50)*

Writing to Pauli in 1919, Max Born expressed the view that the observability principle might hold the key to unravelling the secret of quantum theory:

> I have read your paper in the new issue of *Verhandlungen der Deutschen Physikalishen Gesellschaft* on Weyl's theory with great interest … I have been especially interested in your remark at the end, that you regard the application of the continuum theory, to the interior of the electron as meaningless, because one is then dealing with things that are unobservable in principle. I have pursued exactly such an idea [in quantum theory] for some time, though up to now without positive success.
>
> *(Born to Pauli, 21 December 1919, Pauli, 1979, p. 10 [item 4])*

Throughout the 1920s, Pauli and Born continued to emphasise the importance of the observability principle in physics. As Pauli explained in a letter to Eddington in September 1923, 'the field concept only has a meaning when we specify a reaction, which is in principle possible, by which we can measure the field strength at each point of space-time'. He went on to stress: 'as soon as the reaction ceases to be specifiable or in principle executable the respective field concept is no longer defined' (Pauli to Eddington, 20 September 1923, Pauli, 1979, p. 117 [item 45]). In 1926, Landé recalled that Born had for some years defended the idea that the observability principle held the key for quantum theory. In Göttingen, Born had argued that 'the instantaneous positions and velocities of the electrons' in the atom were '*unobservable in principle*' and should therefore be eliminated from quantum theory altogether (Landé, 1926, p. 455).

It was in conversations with Born and Jordan in Göttingen that Heisenberg was introduced to the key role that the observability principle might play in the further development of quantum theory (Mehra & Rechenberg, 1982b, p. 277–8). In an interview with Kuhn many years later, Heisenberg recalled that 'the idea of having a new theory in terms of observables did indeed originate in Göttingen and was very closely connected with the interest in relativity theory that existed there' (AHQP, 15 February 1963, p. 19). In his 1924 lectures on atomic mechanics, Born spelled out the direction which he felt would ultimately prove fruitful in quantum theory: 'Of these real quantum laws we must require that they involve only observable quantities such as energies, light frequencies, intensities and phases' (Born, 1925, p. 114). The observability criterion was again highlighted in a paper by Born and Jordan in 1925, in which they argued that quantum theory should adhere to the lesson of Einstein's special theory of relativity and admit only relations between 'quantities that are observable in principle' (Born & Jordan, 1925, p. 493). A similar (though, as we shall see later, not identical) emphasis on the observability principle appeared in the introduction to Heisenberg's paper on quantum mechanics in July 1925.

On the surface, the introduction to Heisenberg's 1925 paper on quantum mechanics presented the view that unobservable quantities such as the position of the electron in the atom and the period of its orbit should be excluded from quantum theory. The task of the new quantum mechanics, he maintained, was to construct a non-classical equation of motion for the electron using only observable quantities such as the frequencies and intensities of the radiation emitted by the atom. Drawing on an analogy for the Fourier expansion of position coordinate x in classical theory, Heisenberg substituted infinite Fourier expansions representing the frequencies and intensities of emitted radiation for the classical quantities such as position x in the classical equation of motion $f(x) - x'' = 0$. It was Max Born who first realised that in Heisenberg's mathematical reinterpretation of kinematics and mechanics, the quantum analogue of the position coordinates and the momenta could be represented as matrices. Here was the key that led to the matrix mechanics elaborated by Born and Jordan in the latter part of 1925, and led to the derivation of the well-known non-commutation relation for position and momentum in matrix mechanics $\mathbf{pq} - \mathbf{qp} = \mathbf{h}/(2\pi i)$.

2.2 The renunciation of the electron orbit and the elimination of unobservables

In 1954, Max Born maintained that 'Heisenberg [had] banished the picture of electron orbits with definite radii and periods of rotation because these quantities are not observable' (Born, 1956c, pp. 179–80). This account has been repeated numerous times in the popular literature, yet as a number of historians of science have pointed out, a closer examination of the way in which Heisenberg approached the task of constructing a theory of quantum mechanics shows that this view is mistaken. The observability principle, as it turns out, was not the driving motivation behind the abandonment of the electron orbit. Rather, a reconstruction of the steps that led to Heisenberg's paper demonstrates that he probably inserted it in his paper simply as a *post facto* justification of the new quantum mechanics. Indeed, if we look carefully at the correspondence in the two years preceding Heisenberg's paper on quantum mechanics in 1925, we find that the elimination of the electron orbit in quantum theory had been anticipated, and was in fact the subject of much discussion, particularly by Pauli. The observability principle, however, appears to have played little or no part in these discussions.

A detailed examination of the problems that confronted physicists working on quantum theory between 1920 and 1925 is beyond the scope of this chapter, and in any case the topic has received extensive treatment in the secondary literature (Jammer, 1966; Serwer, 1977; Mehra & Rechenberg, 1982a, 1982b; Hendry, 1984; Darrigol, 1992). Suffice it to say that by 1923, a number of physicists recognised what David Cassidy has termed the 'crisis of quantum theory' (Cassidy, 1976). The difficulty of reconciling the anomalous Zeeman effect with the existing quantum theory and the discrepancy between the calculated orbital frequency of the electron and the frequency of the emitted radiation continued to defy solution by conventional mathematical techniques. 'It becomes increasingly probable', wrote Born in 1923, 'that not only new assumptions will be needed in the sense of physical hypotheses, but that the entire system of concepts in physics will have to be restructured in its foundations' (Born, 1923, p. 542). It was in this atmosphere of 'crisis' that a number of physicists began to suspect that an electron did not move in a circular or elliptical path around the atomic nucleus, as depicted in the Bohr atom.

From as early as 1923, Pauli and Heisenberg had begun to suspect that the difficulties encountered in quantum theory not only pointed to the need for a new dynamics (concept of force), but also called into question classical kinematics (concept of motion) (Heisenberg, 1958b, p. 40). Reflecting on this period in later years, Heisenberg recalled that in their student days in Munich, he and Pauli had become increasingly sceptical of the existence of electron orbits (Heisenberg, 1971, pp. 35–7). In a letter to Sommerfeld in June 1923, Pauli raised the possibility that the dispersion of radiation by the atom was the effect of an oscillation, whose frequency was not associated with the electron orbit (Pauli to Sommerfeld, 6 June 1923, Pauli, 1979, pp. 94–101 [item 37]). Writing to Bohr eight months later, under the weight of mounting difficulties in atomic theory, Pauli again raised what he saw as 'the most important question' confronting quantum theory:

> To what extent is one allowed at all to speak about well-defined trajectories of the electrons in stationary states? I think that this can in no way be assumed as self-evident, especially in view of your observation about the balance of statistical weights in coupling. Heisenberg has in my view hit the mark precisely when he doubts the possibility of determinate trajectories. Doubts of this kind Kramers has never considered as reasonable. I must nevertheless insist upon this, because the point appears to me to be very important.
>
> *(Pauli to Bohr, 21 February 1924, Pauli, 1979, p. 148 [item 56])*

This passage makes it clear that both Pauli and Heisenberg had come to doubt the existence of electron orbits in quantum theory well before Heisenberg's 1925 pivotal paper on quantum mechanics, and significantly, without any reference to the observability principle. In the letter written to Bohr in December 1924, Pauli reiterated this view, insisting that 'not only the dynamic concept of force, but also the kinematic concept of motion of the classical theory shall have to undergo fundamental changes (it is for this reason that I have avoided entirely in my work the designation "orbit")' (Pauli to Bohr, 12 December 1924, Pauli, 1979, pp. 188–9 [item 74]). In his article on quantum theory for the *Handbuch der Physik* completed in 1925 Pauli again stressed that in quantum theory 'it seems that one must renounce the practice of attributing to the electrons in the stationary states, trajectories that are uniquely defined in the sense of ordinary kinematics' (Pauli, 1926, p. 167).

It has become customary to see Pauli's rejection of the orbit as stemming from his commitment to the observability principle in physics. As Henk de Regt puts it: 'Pauli's operationalist view led him to criticize classical kinematical descriptions of atomic processes, especially the concept of electron orbits. He argued that, because the path of an electron cannot be defined operationally, it should have no place in quantum theory' (de Regt, 1999, p. 407). Yet, there is little evidence to suggest that operationalism was behind Pauli's conviction that electron orbits should be eliminated from quantum theory. On the occasions when Pauli spoke about the need to abandon classical kinematics, he gave no indication that his reasoning had anything to do with non-observability of the electron orbit. Rather, the point Pauli repeatedly makes is the difficulty of reconciling the assumption that the electron follows a well-defined trajectory in the atom, with the observed experimental phenomena.

Heisenberg, for his part, was inclined to retain the concept of the orbit in the period between 1923 and 1925 as a heuristic device in an attempt to reconcile the anomalous experimental phenomena with quantum theory, though he stressed that such 'representations have, in principle, only a *symbolic meaning*', and cannot be taken to represent the reality of the electron's motion in the atom (Heisenberg to Pauli, 9 October 1923, Pauli, 1979, pp. 125–7 [item 47]; Heisenberg, 1960a, p. 41). As late as April 1925, Heisenberg would continue to argue that, in spite of difficulties thus far encountered, 'in the present state of quantum theory, one must rely on the use of symbolic, model-like pictures [*symbolische modellmässßige Bilder*] which are formed more or less after the behaviour of electrons in classical theory' – in the sense that they move in well-defined orbits (Heisenberg, 1925b, p. 842).[2] However, by this time he had already recognised the inherent problems with such an approach. Heisenberg's joint paper with Kramers in May 1925 on the optical dispersion of radiation by the atom was perhaps the crucial turning point that persuaded him of the need for a new theoretical approach (Kramers & Heisenberg, 1925). 'This attempt led to a dead end', Heisenberg later recalled, 'but the work helped to convince me of one thing: that one [now] ought to ignore the problem of electron orbits inside the atom' (Heisenberg, 1971, p. 60). On 9 July, he conveyed to Pauli his complete agreement on the need to develop a new basis for kinematics in quantum theory:

> We certainly already agree that the kinematics of quantum theory is totally different from that of classical theory … It is really my conviction that an interpretation of the Rydberg formula in terms of circular and elliptical orbits, according to *classical* geometry, does not have the slightest physical significance and all my wretched efforts are devoted to killing the concept of orbits completely – which cannot be observed anyway – and replace it by a more suitable one.
> *(Heisenberg to Pauli, 9 July 1925, Pauli, 1979, p. 231 [item 96])*

It is instructive to note that while he refers to the fact that the orbits 'cannot be observed anyway', Heisenberg attaches little importance to this point. As the letter makes clear, Heisenberg, convinced that Bohr's orbital model of the hydrogen atom did not have any 'physical significance', was now intent on eliminating the electron orbit from quantum theory altogether. Bohr had repeat-edly emphasised from 1913 that the concept of the stable electron orbit in a stationary state stands in direct violation of classical electrodynamics, but by 1925, Heisenberg had arrived at the view that the problem of describing the electron's motion in quantum theory 'has nothing to do with electrodynamics but is rather – and this seems to me to be particularly important – of a purely *kinematic* nature' (Heisenberg, 1967a, p. 263). This marks a significant

---

[2] Klauss Hentschel has argued that Heisenberg's approach in this paper has much in common with the philosophy of symbolic representation in physics outlined in the introduction in Heinrich Herz's *The Principles of Mechanics* published in 1894 (Hentschel, 1998).

departure from the conceptual starting point of the Bohr–Kramers–Slater paper (Bohr, Kramers & Slater, 1924, p. 795).

The absence of any firm commitment to the observability principle in Heisenberg's published papers and correspondence before summer 1925 is also telling. In an early draft of his paper on quantum mechanics, contained in a letter to Kronig on 5 June 1925, Heisenberg made no mention of the principle (Kronig, 1960, pp. 23–5). By reconstructing Heisenberg's train of thought when developing his pivotal paper, Darrigol (1992) and MacKinnon (1977) have shown that the principle of eliminating unobservables did not play a decisive role. As MacKinnon explains, by the middle of 1925, Born's and Pauli's emphasis on 'the need to restrict quantum theory to observable quantities' had begun to make an impression on Heisenberg, as a result of which he 'accorded this principle a key epistemological role in his formulation though it was incompatible with his actual procedure' (MacKinnon, 1977, p. 163). The observability principle had not *motivated* the decision to eliminate the electron orbit from quantum theory; it had merely served to *justify* it *post facto*.

This idea that the elimination of unobservables played no real guiding role in Heisenberg's thinking leading to his paper on quantum mechanics has been discussed by several historians of science. As Jan Lacki puts it, 'Heisenberg appears to have introduced this requirement [that one should use only observable quantities] *a posteriori*, as a way of organizing and providing methodological unity to his paper, only after having derived its crucial steps' (Lacki, 2002, p. 443). This same idea can be found in the earlier works of Mehra and Rechenberg (1982b, pp. 284–9), Darrigol (1992, pp. 273–4) and MacKinnon (1977, pp. 137, 184–5). Mara Beller has also emphasised that 'the elimination of unobservables was probably invoked *ex post facto* – as justification and not as guiding principle' (Beller, 1983, p. 477). Indeed, as early as 1936, Percy Bridgman suggested that Heisenberg's principle of including only 'intrinsically measurable' quantities was 'formulated after the event as a sort of philosophical justification for its success, rather than having played an indispensable part in the formulation of the theory' (Bridgman, 1936, p. 65).

Heisenberg's approach to physics during the period leading up to his 1925 paper on quantum mechanics had in fact attracted sharp criticism from Pauli. While acknowledging that Heisenberg was a brilliant physicist, perhaps even a genius, Pauli complained to Bohr that Heisenberg was 'unphilosophical' because 'he pays no attention to clear presentation of the basic assumptions and their relationship to previous theories' (Pauli to Bohr, 11 February 1924, Pauli, 1979, p. 143). Indeed, in 1921, Heisenberg had written to Pauli that his own approach to quantum theory was not based on any methodological approach, but was best summed up in the motto 'success justifies the means' [*Der Erfolg heiligt die*

Mittel] (Heisenberg to Pauli, 19 November 1921, Pauli, 1979, p. 38). In Cassidy's words, 'in its most creative form Heisenberg's style of physics was characterized by paradox, inconsistency, and pragmatism' (Cassidy, 1979, p. 189).

The stated aim of the 1925 paper was to abandon the concept of the orbit, and in its place set up a system of mathematical equations which used the observable frequencies and intensities of the radiation emitted by the atom during state transitions. By abandoning the idea that the electron moved in a continuous trajectory in space and time, Heisenberg was confronted with the problem of how the new quantum mechanics could be extended to describe the observed tracks left by the electron in the Wilson cloud chamber, which appeared to show that *free* electrons in non-stationary states move in well-defined trajectories. As Heisenberg would later put it, his strategy 'was to relinquish at the first the concept of electron paths altogether, despite its substantiation by Wilson's experiments', and then to attempt to discover 'how much of the electron-path concept can be carried over into quantum mechanics' (Heisenberg, 1965, p. 292). This was a problem Heisenberg would need to resolve before he could claim to have properly developed a complete theory of quantum mechanics. It was therefore necessary for him to justify his renunciation of the electron orbit. However, in resorting to its non-observability, Heisenberg ran into problems. To this extent, the concept of observability in his paper requires closer examination.

2.3 Heisenberg's version of the observability principle

While historians and philosophers of science including Lacki, MacKinnon and Darrigol have devoted their attention to a historical reconstruction of the strategy Heisenberg actually employed in his paper on quantum mechanics, little attention has been devoted to the manner in which Heisenberg articulated the principle of observability in his 1925 paper. Yet, when one looks closely at the paper, it is clear that Heisenberg was rather tentative about committing himself to the view that the electron orbit is in fact unobservable in principle. There are two dimensions to this. First, Heisenberg did not argue, as did many of his contemporaries, that a physical theory *must* only use observable quantities. His contention was simply that given the difficulties already encountered in quantum theory, it 'would seem *more reasonable*' to restrict oneself to 'relations between observable quantities' (Heisenberg, 1967a, p. 262, emphasis added). And secondly, Heisenberg was equivocal on the question of whether the electron's position and period of orbit could be considered unobservable *in principle*. I will address each of these in turn.

Heisenberg never makes the strong claim that only observable quantities should be used in constructing a theory in physics. Indeed, in the final paragraph of the paper, Heisenberg conceded that it remained to be seen 'whether a method to determine quantum-theoretical data using relations between observable quantities, such as that proposed here, can be regarded as satisfactory in principle'. Such a method, he admitted, might indeed constitute 'far too rough an approach to the physical problem of constructing a theoretical quantum mechanics' (Heisenberg, 1967a, p. 276). In the introductory section of the paper, Heisenberg had argued that Bohr's quantum theory had not resolved the 'fundamental difficulties' which 'arise in the problem of "crossed fields" (hydrogen atom in electric and magnetic fields of differing directions)'. Moreover, 'the extension of quantum rules to the treatment of atoms having several electrons has proved unfeasible' (Heisenberg, 1967a, p. 261). He thus drew the conclusion that 'the *Einstein–Bohr* frequency condition (which is valid in all cases) already represents such a complete departure from classical mechanics' or more precisely 'from the kinematics underlying this mechanics' that we can no longer maintain the idea that the electron moves along a well-defined orbit in the atom (Heisenberg, 1967a, p. 262). Heisenberg wrote:

> In this situation it seems sensible to discard all hope of observing hitherto unobservable quantities, such as the position and period of the electron, and to concede the partial agreement of the [old] quantum rules with experience is more or less fortuitous. Instead it would seem more reasonable to try to establish a theoretical quantum mechanics, analogous to classical mechanics, but one in which only relations between observable quantities occur.
>
> *(Heisenberg, 1967a, p. 262)*

In the Born–Heisenberg–Jordan paper on matrix mechanics written in November 1925, the observability principle was again presented as having played a heuristic role (Born, Heisenberg & Jordan, 1926). What is important here is that nowhere do the authors attempt to argue that physics *must* use only observable quantities. This point is often missed by historians and philosophers, who are quick to read Heisenberg as holding an *epistemological* commitment to positivism. Daniel Serwer has rightly pointed out, for Pauli, 'the restriction to observables' was 'a categorical imperative', whereas for Heisenberg it was 'merely another direction of inquiry' (Serwer, 1977, p. 245). The idea of using observable quantities was presented by Heisenberg as a strategy which had been resorted to when confronted with the difficulties that had plagued quantum theory up until that time. Ironically though, as we now know, it actually played no role in the development of the theory.

However, there is another sense in which Heisenberg tempered his appeal to the principle of observability. If we read the introduction closely, we find

Heisenberg was well aware that it was by no means a straightforward matter to regard the electron's position and period of its orbit as unobservable *in principle*. Indeed, in the introduction to the paper, Heisenberg wavers between characterising the electron's position and the period of its orbit as '*bis jetzt unbeobachtbar*' or '*bisher unbeobachtbar*' (hitherto unobservable), and *prinzipiell unbeobachtbar* (unobservable in principle) (Heisenberg, 1925a, pp. 879–81). In the passage quoted earlier, Heisenberg insisted that we should 'discard all hope of observing hitherto unobservable quantities' such as the position and period of the electron, not because of the impossibility of measuring them using current experimental techniques, but because of the theoretical problems associated with the inclusion of such quantities into quantum theory. Note, strictly speaking this is not just an argument for the *elimination* of such quantities, but more importantly it is an argument for treating such quantities as *unobservable in principle*.[3]

We find further evidence of this line of argument in other passages in the introduction to the 1925 paper. There, Heisenberg explains that it is precisely because of the problems encountered in quantum theory thus far that one could safely abandon all hope of observing the position and period of the electron in the atom. He proposed instead that in constructing a new quantum mechanics, we should focus exclusively on such quantities as the intensities and frequencies of the emitted radiation that had *actually been observed*. In the opening paragraph, Heisenberg reflected on the use of the Bohr quantisation rules, which were interpreted as referring to the 'hidden' electron orbits. Here it is worth quoting Heisenberg at length:

> It is well known that the formal rules, which are used in quantum theory for calculating observable quantities such as the energy of the hydrogen atom, may be seriously criticized on the grounds that they contain, as a basic element, relationships between quantities that are *apparently* unobservable in principle, e.g. position and period of revolution of the electron. Thus these rules lack an evident physical foundation, unless one wants to retain the hope that the *hitherto unobservable quantities may later come within the realm of experimental determination*. This hope might be regarded as justified if the above-mentioned rules were internally consistent and applicable to a clearly defined range of quantum mechanical problems.
>
> *(Heisenberg, 1967a, p. 261, emphasis added)*

---

[3] Few historians have noted the significance of this point in Heisenberg's paper. Max Jammer, a notable exception, has pointed out that Heisenberg did acknowledge in his paper that we must consider the possibility 'that future progress in experimental techniques would eventually make it possible to measure these quantities'. To this extent, Jammer contends that 'Heisenberg's rejection of these quantities as unobservable was based on two empirical facts, the experimental impossibility of directly measuring them *and* the practical failure of a theory which assumed them to be observable' (Jammer, 1966, p. 199).

That Heisenberg should initially describe the position and period of the electron's orbit as 'apparently unobservable in principle' is highly revealing. Here he acknowledges that such quantities are 'hitherto unobservable' given that current experimental techniques did not *yet* allow physicists to directly observe the electron's position or the period of its orbit. While he was careful to point out one might hold out the hope that such 'hitherto unobservable quantities may later come within the realm of experimental determination', he dismisses such a possibility as unlikely, given that 'only the hydrogen atom and its Stark effect are amenable' to the existing quantum rules. Put simply, Heisenberg proposed that one should abandon 'all hope of observing hitherto unobservable quantities such as period and position of the electron', not because they remained beyond the reach of any conceivable measuring techniques – as Mach and Einstein had argued in the case of absolute time and space – but rather because all attempts to explain atomic phenomena through the concept of the classical periodic motion of the electron had up until now failed. In arguing this way, Heisenberg adopted a brazenly pragmatic attitude not only to the employment of the observability principle, but to the very meaning of the concept of observability itself.

The problem of how to define 'observability' was never adequately resolved by physicists in the 1920s. In a letter to Eddington in 1923 Pauli pointed out that 'we do not have to concern ourselves with technical difficulties' when considering the epistemological question of whether a quantity is observable *in principle* in physics (Pauli to Eddington, 20 September 1923, Pauli, 1979, p. 117 [item 45]). Max Born, who was keen to draw attention to the significance of the observability principle for both relativity and quantum mechanics, also appears to have recognised the problem of distinguishing between quantities that were 'hitherto unobservable' and those that were 'unobservable in principle'. In his 1925–6 lectures on atomic dynamics at MIT, Born had attempted to defend the view that the electron's path inside the atom was unobservable *in principle*, much as Pauli had done for the concept of the electric field strength in the interior of the electron:

> [N]o one has been able to give a method for the determination of the period of an electron in its orbit or even the position of an electron at a given instant. There seems to be no hope that this will ever become possible, for in order to determine lengths or times, measuring rods or clocks are required. The latter, however, consist themselves of atoms, and therefore breakdown in the realm of atomic dimensions … At this stage it appears justified to give up altogether the description of atoms by means of such quantities as "coordinates of the electrons" at a given time, and instead utilize such magnitudes as are really observable.
>
> *(Born, 1960, p. 69)*

Although Heisenberg had avoided making any direct reference to the experimental techniques which might be employed in the determination of such quantities in his 1925 paper, Born attempted to show why it was not possible to observe the position of an electron in the atom. As Mara Beller explains: 'The argument was that because there are no measuring devices for observing intra-atomic orbits or for measuring the positions and periods of revolution of electrons, the concepts of time and space cannot be applied to the realm of atomic dimensions' (Beller, 1988, p. 151). Yet, the passage quoted above shows that Born was less than completely convinced. His use of the words, 'At this stage it appears justified' suggests that he was somewhat equivocal about whether the coordinates of electrons could legitimately be considered unobservable *in principle*. Indeed, as Heisenberg had already realised, the fact that one could not as yet observe the electron's position in the atom did not necessarily mean that it would remain impossible in the future. In fact, Born described 'the position, velocity and period of the electron', not as unobservable *in principle*, but rather as 'magnitudes of very *doubtful* observability' (Born, 1960, pp. 68–9).

Born would continue to exercise caution in his pronouncements about non-observability. In his Oxford lecture, in August 1926, he argued: 'Of course it is not forbidden to believe in the existence of these co-ordinates; but they will only be of physical significance when methods have been devised for their experimental observation' (Born, 1927, p. 356). Born's equivocations serve to highlight the shaky ground on which his argument for the elimination of the electron orbit rested. If we are to trust the account given to us by Heisenberg in his interview with Kuhn many years later, other physicists had already begun to question the view that the electron's orbit was really unobservable *in principle*. According to Heisenberg, in discussions in Göttingen around 1925–6, Burckhardt Drude raised the possibility that the electron orbit might well become amenable to observation with the future refinement of experimental techniques. In his interview with Kuhn in 1963, Heisenberg recalled his conversation with Drude:

> He started arguing with me about quantum mechanics. I always said that there are no electronic orbits – the whole thing is more abstract. He came more from the experimental side and he was younger and was still entirely in the classical ideas. He told me, "Well, I don't believe a word of your story about non-existing electronic orbits. First of all, the electron comes out, and therefore it must have been in the atom. Then you say there is no orbit. But if you would take a very good microscope, for instance a gamma ray microscope, why shouldn't it be possible? Then you could see the orbit". Now this remark made me feel terrible. I thought, "Well, after all, it's true. The gamma ray microscope has the resolving power which should do it".
>
> *(AHQP, 22 February 1963, p. 26)*

It seems that the problem of whether or not the electron orbit should play any role in quantum theory was for many physicists in Göttingen inextricably linked to the question of whether an orbit was really observable *in principle*. That is to say, it was unclear whether it was conceivable that the period of the electron orbit could become amenable to experimental determination with the refinement of measurement techniques. Ironically, Drude's suggestion of a gamma-ray microscope would later prove crucial to Heisenberg's thought-experiment when discussing the measurement of the position of an electron in the 1927 paper on the uncertainty relations. Drude, for his part, defended the existence of electron orbits by recourse to a thought-experiment demonstrating the theoretical possibility of observing the electron orbit in the atom. For physicists committed to the opposing view, the matter was decidedly more difficult. However, as we have already seen, considerations of this kind did not prove decisive in the formal development of Heisenberg's theory of quantum mechanics. While Drude's remarks were somewhat disconcerting for Heisenberg from a philosophical point of view, they ultimately did not deter him from the view that in the atom 'there are no electronic orbits'.

2.4　The meaning of observability: a discussion with Einstein

Our analysis so far has concentrated on whether or not the *electron's position* and *period of orbit* were quantities that were unobservable in principle. However, the concept of observability more generally was the subject of careful reflection around the time of Heisenberg's paper. Aage Peterson, in *Quantum Mechanics and the Philosophical Tradition* observes: 'The idea of observability went through a significant transformation in the period between Heisenberg's fundamental paper on the quantum formalism [in 1925] and his paper on its physical interpretation [in 1927]' (Petersen, 1968, p. 95). Physicists and philosophers alike had deemed the concept of observability problematic well before it surfaced in the specific context of quantum mechanics. In 1919, Hermann Weyl defended his program of a unified field theory against Pauli's criticisms, which we examined earlier, by arguing that contrary to Pauli's claim it was possible 'to measure the fields in the interior of the electron' provided that the effects of the fields 'grow to an immediately noticeable magnitude' (Weyl to Pauli, 9 December 1919, Pauli, 1979, p. 6 [item 2]). In discussions on this very issue, following Weyl's address in 1920 in Bad Nauheim at the Assembly of German Scientists and Physicians, Einstein commented that the notion of observability was a far more complicated notion that is usually thought (Weyl, 1920, p. 651). Heisenberg's introduction of the principle of observability must be situated against this intellectual backdrop.

While a number of physicists questioned the role of observability in the new theory, the most decisive blow against the idea that observability should play a major epistemological role in the new quantum mechanics came from a discussion Heisenberg had with Einstein some time in April 1926. At Einstein's invitation, Heisenberg travelled to Berlin to discuss the new quantum mechanics, and their conversation soon turned to the concept of observability.[4] To Heisenberg's surprise, Einstein was quite critical of the idea that one should include only observable quantities in a physical theory. Far from advocating the viewpoint, which Heisenberg had assumed to have underpinned Einstein's special theory of relativity, Einstein argued instead: 'it may be useful to keep in mind what one has actually observed' but 'it is wrong to try to formulate a theory on observable magnitudes alone. In reality the opposite happens. It is the theory which decides what we can observe' (Heisenberg, 1971, pp. 63–4). The claim that the electron orbit is unobservable in principle, Einstein explained, 'is therefore an assumption about a property of the theory that you are trying to formulate'. In his interview with Kuhn, Heisenberg recalled his discussion with Einstein:

> He had explained to me that what is observed or not is decided by the theory. Only when you have the complete theory can you say what can be observed. The word observation means that you do something which is consistent with the known physical laws. So long as you have no laws in physics you don't observe anything. Well, you have impressions and you have something on your photographic plate, but you have no way of going from the plate to the atoms. If you have no way of going from the plate to the atoms, what is the use of the plate? That was an argument which made a strong impression on me – that was the discussion with Einstein.
>
> *(AQHP, 25 February 1963, p. 19)*

This view would exert a decisive influence on Heisenberg. In the 1925 paper, Heisenberg had assumed that it was possible to draw a clear distinction between quantities such as the electron's position and period of its orbit, which he regarded as unobservable, and the quantities such as the frequencies and intensities of the emitted radiation, which were observable. As Einstein now explained, prior to the establishment of a theory, it is not possible to draw a clear distinction between observable and theoretical quantities, and to this extent one could not categorically define the electron's position and period of its orbit as unobservable *in principle*.

---

[4] We know the meeting took place on 26 April 1926, from Heisenberg's letter to his parents dated 29 April. However, no documentary evidence exists of what was discussed during this historic encounter (Mehra & Rechenberg, 2000, pp. 131–2). This conversation was referred to in Heisenberg's later recollections and was reconstructed in some detail in *Physics and Beyond*. David Cassidy has offered a slightly different reconstruction of their discussion (Cassidy, 1992, pp. 237–9).

Einstein's argument can be illustrated by considering the example of the observation of the charge of an electron. According to classical theory, we may empirically determine the electron's charge by observing the curvature of its path in an electric field in the Wilson cloud chamber. If we know the momentum of the electron at any given time, as well as the strength of the electric field, we can calculate the charge of the electron. One might claim to have 'measured' the charge of the electron, even though what we actually 'observe' is the radius of the curvature in the electron's path. The observability of the electron's charge ultimately depends on the validity of classical electrodynamics, and the system of laws and concepts employed in that theory.

Heisenberg's writings after his discussion with Einstein reflect his awareness of the problem posed by the concept of observability. Though it did not bring to an end his flirtation with operationalism, as evidenced by his attempt to give the concepts of position and velocity an operational meaning in 1927, he did express some degree of confusion about the observability issue. Indeed, as we shall see in Chapter 5, the operational point of view, which characterises the attempts to define concepts in the 1927 paper, was short-lived. By the second half of 1927, it had become clear to Heisenberg that the observability principle could not serve as the foundation on which to construct a physical theory. In the introduction to the report presented by Born and Heisenberg at the Solvay conference in Brussels in October 1927, the authors stated that it had been their intention 'to introduce new notions' into quantum mechanics 'by a precise analysis of that which is "essentially observable"'. However, as he and Born now explained:

> This is not tantamount to establishing the principle that it is possible and even necessary to draw a neat separation between that which is 'observable' and that which is 'unobservable'. Once a system of concepts is given one can ascribe observations to other facts that, properly speaking, are not directly observable, and the boundary between that which is observable and that which is not observable, becomes entirely undetermined. But when a system of concepts is itself still unknown [as was the case in quantum theory in 1925], quite naturally one is interested only in the observations themselves, without drawing conclusions [i.e. that the electron moves in a well-defined orbit], because otherwise false ideas and old prejudices would thwart the understanding of physical relations.
>
> *(Born & Heisenberg, 1928, pp. 143–4)*

This passage reflects a rather more subtle position with regard to observability in physics than we find in the introduction to the 1925 paper. In refusing to draw a sharp demarcation between the observable and the unobservable within a particular physical theory, Heisenberg here departed from the view then held by many of the logical positivists (e.g. Carnap, 1975). What we can regard as 'observable' in quantum mechanics depends on the structure of the theory itself.

We should not, of course, overlook the fact that earlier that same year, Heisenberg had again attempted to use observability arguments in an explicit attempt to operationally define the concept of position and velocity in quantum mechanics using an imaginary gamma-ray microscope. Yet, as we shall see later, after extensive discussions with Bohr in Copenhagen, Heisenberg abandoned the operationalist standpoint which characterises the 1927 paper in the months that followed. To this extent, Heisenberg's early forays into 'operationalism' in his seminal papers on quantum mechanics in the 1920s were short-lived, and, as we shall see later, by the 1930s his thought had clearly moved in other directions.

What is important to note is that Heisenberg's conception of observability clearly undergoes a shift after 1925, as a result of his discussions with other physicists, in particular, Einstein (and perhaps Drude). As he was to put it in 1968, 'when one has invented a new scheme which concerns certain observable quantities, then of course the decisive question is: which of the old concepts can you really abandon?' (Heisenberg, 1968, p. 37). By the second half of 1927, he was no longer arguing that only what is observable in principle should be included in quantum mechanics. His position now was that in times of conceptual difficulty, it may be useful to confine oneself to quantities that have *actually been observed*, in the interests of establishing a fully consistent and empirically adequate theory. Indeed, pragmatic considerations of this kind appear to have guided his work on the S-matrix theory of elementary particles in the 1940s.

Writing to Wien in June 1926, Schrödinger complained that 'all philosophy about "non-observability in principle"' had simply confused the issue (Schrödinger to Wien, 18 June 1926, Mehra & Rechenberg, 2000, p. 137). Indeed, Heisenberg might well have agreed with this assessment, given his later critical reflections over the problem of observability in quantum theory. It is significant that in his critique of Schrödinger's papers on wave mechanics which appeared in 1926, Heisenberg did not take Schrödinger to task for introducing *unobservable* quantities such as the ψ-function. In an interview with Kuhn, Heisenberg freely admitted: 'Even in quantum mechanics, one can say that finally one did not work with observable quantities because one actually had to introduce – and did introduce – the Schrödinger wave function' (AHQP, 15 February 1963, p. 21).

Heisenberg's critique of the concept of observability reveals both his own pragmatic attitude to theory construction, and his capacity and predisposition to reflect on the philosophical foundations of physics. But it also tells us much about Heisenberg's early attitude to positivism. By the late 1920s, positivism had exercised a deep influence on many physicists, largely because of its perceived influence on Einstein's theory of relativity and the rise to prominence of the Vienna Circle. While initially hopeful that such a viewpoint would buttress the new theory of quantum mechanics, Heisenberg would become increasingly sceptical of such an approach. At the very least, Heisenberg's equivocations about observability reveal his uneasiness with the idea that observability, and an empiricist philosophy of physics more generally, could provide a foundation for quantum mechanics. This is exemplified in his critique of the notion of observability principle after 1925, but it also emerges much more forcefully in other ways which will be examined in more detail later on, particularly in Chapter 5, where I examine his analysis of indeterminacy.

By 1925, it had became clear to Heisenberg that one could no longer interpret the electron as moving in a well-defined orbit according to classical kinematics. But Heisenberg took this further – as we shall see in the next chapter, he argued that it was not possible to describe the electron's motion using the usual concepts of space and time. This presented Heisenberg with the dilemma of making physical sense of the new theory. In particular, it was not clear to what extent one could speak of the 'motion' of an electron at all. The debate between Heisenberg and Schrödinger over the physical significance of wave mechanics would illuminate the different philosophical approaches that underpinned the attempt to interpret the new physics. Heisenberg's own philosophical approach to the interpretation problem and his views on what is meant by *understanding* in physics must be seen in the context of his debate with Schrödinger in 1926. But, as I shall argue in the next chapter, his views were also reflective of the empiricist reaction to Kantian philosophy which had emerged in the German-speaking world in the second half of the nineteenth century, and which continued to gather momentum in the early decades of the twentieth century. Yet, as we shall see, in spite of the influence of these empiricist currents, Heisenberg retained something of a realist outlook in his attitude to quantum mechanics.

3

The problem of interpretation

Matrix mechanics posed a formidable challenge for physicists in understanding the quantum world. While the new mathematical scheme undoubtedly signified a fundamental advance over the previous methods associated with Bohr's theory of the atom, it remained unclear how to interpret the abstract equations of motion in which the coordinates and momenta of the electrons were represented by infinite matrices. The appearance of a series of papers on wave mechanics by the Austrian physicist Erwin Schrödinger in early 1926 suggested an altogether different approach to the interpretation problem – electrons are not particles but rather waves. This led to a vigorous debate between Heisenberg and Schrödinger on the physical meaning of the new wave mechanics. A close examination of this debate illuminates the underlying philosophical issues at stake for the two protagonists, but also serves to show how Heisenberg's view was shaped by his critical dialogue with Schrödinger and the empiricist reaction to Kant in the 1920s.

In this chapter, I explore the contrasting philosophical viewpoints that underpinned the two approaches. Schrödinger was convinced that in order to 'understand' quantum theory, it must be possible to visualise the inner workings of the atom in space and time. Heisenberg and Bohr argued that in quantum mechanics we must abandon the 'classical ideal of understanding', and instead we must learn what the word 'understanding' means in this new context. In contrast to Bohr, however, Heisenberg sought to define this term along instrumentalist lines. Here he appealed to the transformation in our understanding of space and time brought about by Einstein's general theory of relativity. According to Heisenberg, we may justifiably claim to 'intuitively understand' a theory (such as Einstein's theory of curved three-dimensional space) when all experimental consequences of the theory are clear to us, and the theory is internally consistent, that is, it contains no contradictions. For Heisenberg, the task of interpretation was therefore *not* to interpret quantum mechanics *classically* (as a wave or particle theory), but rather to reinterpret the classical analytic concept of motion *quantum mechanically*. Once

Heisenberg realised that the tracks left by the electron in the Wilson cloud chamber, which appeared to show continuous well-defined particle trajectories, could be re-interpreted quantum-mechanically, as a series of discontinuous interactions between the electron and the molecules of water vapour in the cloud chamber, he considered that the problem of interpretation had effectively been solved.

In his clash with Schrödinger in 1926, Heisenberg claimed that the behaviour of the electron in the atom was *unanschaulich* or unvisualisable, as its motion could not be represented in space and time, but in his 1927 paper on the uncertainty relations, he would argue that quantum mechanics was in fact an *anschaulich* or intuitive theory. This shift in Heisenberg's perspective turns on his subtle redefinition of the word *Anschaulichkeit*, which was closely connected with his positivistic conception of 'understanding'. While historians have correctly pointed out that Heisenberg's instrumentalist redefinition of 'visualisation' signifies a critical response to Schrödinger's emphasis on 'classical' space-time visualisation, they have largely ignored the historical background to Heisenberg's view of *Anschaulichkeit*, which finds its expression in an empiricist strand of post-Kantian German philosophy originating with Helmholtz in the second half of the nineteenth century. The emergence of non-Euclidean geometry, both in mathematics and in physics, gave rise to a new notion of *Anschaulichkeit* in the late nineteenth and early twentieth centuries, forming the intellectual context in which we should situate Heisenberg's redefinition of the term in his 1927 paper.

While Heisenberg appears to have been largely influenced by the positivist currents of his time, certain passages in his writings reveal a certain disposition towards some form of realism. Indeed, in Heisenberg's later writings this is more prominent. We can make sense of Heisenberg's seemingly conflicting views by distinguishing between the *nature* of an entity, like an electron, and the *structure* of a theory, like quantum mechanics. For Heisenberg, it is the latter, captured by the mathematical equations of quantum mechanics, which alone gives us an insight into the mathematical *form* of reality itself. In the terminology used in contemporary philosophy of science, we can say that Heisenberg was therefore closer to 'structural realism', a position which can be traced back to such thinkers as Henri Poincaré, Ernst Cassirer and Moritz Schlick in the early part of the twentieth century. By reading Heisenberg in this way, we can make better sense of many of the more enigmatic passages and apparent contradictions in Heisenberg's later writings.

3.1 Reconceptualising the electron's motion

The emergence of matrix mechanics in 1925 in Göttingen raised new and difficult problems regarding how to physically interpret the new theory. The classical analytical conception of motion, understood as a continuous curve in the space of ordinary geometry, was replaced by a quite different formalism. But it was as yet unclear what the equations of matrix mechanics meant physically. In an unpublished manuscript entitled 'The Problem of Space in the New Quantum Mechanics', written some time around the middle of 1926, Hans Reichenbach pointed to what he saw as the fundamental problem of matrix mechanics, as it had been developed in Göttingen under Heisenberg, Born and Jordan:

> Yet if the authors believed that they had furnished at the same time a new kinematics of the atom, it must be pointed out that their view was hardly justified. Kinematics means the presentation of motion processes in space; but an interpretation of matrix mechanics had not even been attempted ... We must never forget that a renunciation of the planetary character of the electron system is not equivalent to a renunciation of a spatial interpretation.
>
> *(Reichenbach, 1991, p. 33)*

By the second half of 1925, Heisenberg had abandoned all hope of giving a classical interpretation of the electron's motion in space and time. In November 1925, Born, Heisenberg and Jordan emphasised that the new quantum mechanics 'would labour under the disadvantage of not being directly amenable to geometrically visualisable interpretation, since the motion of electrons cannot be described in terms of the familiar concepts of space and time' (Born, Heisenberg & Jordan, 1967, p. 322). But far from regarding this as a weakness in understanding the atomic world, Heisenberg felt that the abandonment of a visualisable space-time description of the electron's motion made possible by matrix theory constituted a great step forward. By the same token, he acknowledged that a new *concept of motion* was not yet at hand. Matrix mechanics had as yet no solution to the problem of describing the motion of a free particle, which appeared to follow a well-defined trajectory in the Wilson cloud chamber.

Yet, it is a mistake to suggest that Heisenberg was uninterested in an interpretation of matrix mechanics. In a letter to Edward MacKinnon, Heisenberg took issue with the suggestion that he had remained satisfied with a purely formal description of quantum mechanics: 'You express the opinion that only Bohr felt the need for a physical interpretation of quantum mechanics while the other physicists including myself considered the mathematical scheme as a sufficient explanation of the phenomena. I believe that I have always shared the opinion of Bohr and disagreed with the other physicists' (Heisenberg to MacKinnon, 12 July 1974, MacKinnon, 1977, p. 187). This is one case where Heisenberg's later recollections are confirmed by an examination of his published writings and correspondence of the period. Heisenberg's correspondence in the second half of 1925 reveals that, contrary to what is sometimes maintained, he was actively interested in finding a physical interpretation of the new matrix theory.

Perhaps the earliest evidence of a concern for the physical interpretation can be found in a letter to Pauli in June 1925, where Heisenberg admitted that he was troubled by the question of 'what the equations of motion really mean, when one treats them as relations between transition probabilities' (Heisenberg to Pauli, 24 June 1925, Pauli, 1979, p. 228 [item 93]). Writing to Bohr in October of that year, Heisenberg pointed out that he was still unhappy that a physical interpretation of the theory had continued to elude physicists in Göttingen. Although pleased that the treatment of atomic spectra by the methods of matrix mechanics was so 'mathematically elegant', he lamented that 'we cannot yet find any physical meaning of it' (Heisenberg to Bohr, 21 October 1925, Bohr, 1984, p. 368). In the joint paper with Born and Jordan on matrix mechanics, Heisenberg remained hopeful that a physical meaning of the *Bewegungsproblem* [problem of motion] could still be found in the abstract equations of matrix theory, perhaps through new 'quantum' concepts of space and time (Born, Heisenberg & Jordan, 1926, p. 559).[1] This view is again expressed in the concluding section of his paper submitted in December 1925, where Heisenberg acknowledged that the 'real geometric and kinematic meaning' of the equations of matrix mechanics 'has not yet been made entirely clear' (Heisenberg, 1926d, p. 705). In a letter to Dirac in November 1925 Heisenberg emphasised that he was not content with merely stating that in quantum mechanics, 'the mathematical operations which are used to deduce the physical results are different from the classical theory'. Instead, Heisenberg insisted that in quantum theory 'we really have to do with a change in the kinematics'. To this extent, Heisenberg remained convinced there was 'an important physical point' in the new theory, which had not yet been fully elucidated (Heisenberg to Dirac, 23 November 1925, Dirac, 1977, p. 126). Indeed, just a week earlier, Heisenberg had written to Pauli, complaining of the attitude in Göttingen, which seemed to completely ignore the physical side of the problem (Heisenberg to Pauli, 16 November 1925, Pauli, 1979, pp. 255–7 [item 105]). The interpretation of the new mathematical scheme of quantum mechanics now constituted a task of critical importance.

However, precisely what Heisenberg meant by an *interpretation* is the crux of the issue here. Having abandoned the possibility of giving a classical trajectory

---

[1] In the English translation of the paper by van der Waerden, the term *Bewegungsproblem* is translated as 'problem of dynamics' (Born, Heisenberg & Jordan, 1967, p. 323). However, 'problem of motion' is a more accurate translation, given Heisenberg's stated aim to develop a new quantum kinematics.

description of the electron's motion in the new quantum mechanics, it remained unclear to Heisenberg precisely how the physicist could describe the behaviour of the electron in the atom. While a full treatment of this question requires that we examine Heisenberg's attempt to redefine the concepts of space and time in quantum theory – which I focus on in Chapter 5 – it is important first to look at Heisenberg's critical reaction to other attempts to provide a physical interpretation of quantum mechanics. It was precisely this problem that the Austrian physicist Erwin Schrödinger took up in the first half of 1926. As we shall see, the debate between Heisenberg and Schrödinger on this question was to become one of the defining clashes of the 1920s and brought out the different philosophical attitudes regarding what it means to give a 'physical interpretation' of quantum mechanics.

3.2 The physical meaning of Schrödinger's wave mechanics

An entirely different mathematical approach to quantum theory was developed by the Austrian physicist Erwin Schrödinger in the first half of 1926. Like Heisenberg, Schrödinger sought to formulate a new mechanics and kinematics for the electron, but unlike the matrix-theoretical approach developed in Göttingen, which was based on the 'quantisation' of the classical equation of motion for a point-particle, Schrödinger took as his point of departure the wave equation to describe the electron in the stationary state of a hydrogen atom. Schrödinger hoped that by treating the quantisation of periodic motion as an eigenvalue problem, one could shed new light on the physical reality of the electron in the atom. As he put it in his first paper on wave mechanics: 'we should try to connect the function ψ with some *vibration process* in the atom which would more nearly approach the reality than the electron orbits, the real existence of which are very much questioned today' (Schrödinger, 1928c, p. 9). Indeed, Heisenberg was initially enthusiastic about the possibilities opened up by Schrödinger's theory. Writing to Dirac in April 1926, he remarked that Schrödinger's approach was 'closely connected with [matrix] quantum mechanics' and moreover that Schrödinger's treatment of the hydrogen atom constituted an important advance from which one could 'win a great deal for the physical significance of the theory' (Heisenberg to Dirac, 9 April 1926, Dirac, 1977, p. 131). Writing to van der Waerden in later years, Heisenberg recalled that like Schrödinger, he had formed the view that 'something must oscillate in the atom with this right frequency and that must mean that one must introduce some strange new kinematics for the electron' (Heisenberg to van der Waerden, 8 October 1963, van der Waerden, 1967, p. 29).

While Schrödinger's discovery of the wave equation had aroused Heisenberg's interest in the spring of 1926, Heisenberg strongly disapproved of the physical interpretation of wave mechanics pursued by Schrödinger in his early papers. Schrödinger had proposed that rather than imagine the electron as a particle orbiting the nucleus, we should instead conceive of the stationary state of an atom as a standing harmonic matter wave surrounding the nucleus of the atom. According to Schrödinger's wave theory of matter, 'the charge of the electron is not concentrated in a point, but is spread through the whole of space' (Schrödinger, 1926, p. 1066). In this way, Schrödinger proposed that the quantisation of energy in atomic theory could be explained away as an eigenvalue problem in the solution of a fundamental wave equation. In a letter to Dirac in May 1926, Heisenberg remarked: 'I quite agree with your criticism of Schrödinger's paper with regard to the wave theory of matter. This must be inconsistent just like the wave theory of light' (Heisenberg to Dirac, 26 May 1926, Mehra & Rechenberg, 2000, p. 202). It seemed to Heisenberg that in attempting to develop a continuous wave theory of matter, Schrödinger had simply avoided the fundamental element of discontinuity that was so typical of all quantum phenomena. As he later told Kuhn:

> [M]y disappointment about Schrödinger's first paper came from this very point that he thought he could give an interpretation in the old sense. I thought that such an interpretation in the old sense does not exist. That was my very deep conviction. So I could not put it into quite rational terms. That could only be done a year later. But I felt that if a man tries to make such a direct interpretation – that he says that an electron is a wave, and so on – then he must certainly be wrong. That was the only thing I was convinced of.
>
> *(AHQP, 22 February 1963)*

Heisenberg's critical reaction to Schrödinger's efforts to develop wave mechanics was guided to a large extent by the 'deep conviction' he had acquired in 1924–5 regarding the kind of reality appropriate to electrons in quantum theory. Matrix mechanics and wave mechanics emerged from entirely different starting points and were based on differing views of the status of energy quantisation or discontinuity in the atom. In spite of the different ontological viewpoints expressed by Heisenberg and Schrödinger, in 1926 the latter published a paper in which he demonstrated the formal equivalence of the two theories, though precisely in what sense the two theories were equivalent is still the subject of some debate (Schrödinger, 1928b).[2] Pauli also outlined

---

[2] Muller (1997) has argued that the equivalence of matrix mechanics and wave mechanics is one of the founding myths of quantum theory. More recently, Slobodan Perovic has also argued that it was never Schrödinger's intention to demonstrate the formal isomorphism of matrix mechanics and wave mechanics in his 1926 paper, but merely to show that the two theories give equivalent results for the energy states of the Bohr atom (Perovic, 2008).

his own derivation of the equivalence of matrix mechanics and wave mechanics in a letter to Jordan, dated 12 April 1926 (van der Waerden, 1973, pp. 278–82). Above all, the physical interpretation of the new theory continued to arouse heated debate between the two camps. Writing to Pauli in June 1926, Heisenberg expressed his intense dislike for the wave theory of matter: 'The more I ponder on the physical part of Schrödinger's theory the more detestable I find it' (Heisenberg to Pauli, 8 June 1926, Pauli, 1979, p. 328 [item 136]). Matrix theory, with its renunciation of the classical visualisability, in Heisenberg's view, offered a more secure path to understanding the quantum world.

Heisenberg cited two main reasons for his critical reaction to Schrödinger's wave interpretation. First, it was not possible to account for certain experimental phenomena using a wave theory of matter, and secondly, Schrödinger's wave equation, far from describing a wave in three-dimensional space, was an abstract differential equation in multi-dimensional configuration space. In the first case, Heisenberg argued that it was impossible to account for a number of experimental results including Planck's radiation law and Compton scattering by using a continuous wave theory of matter to account for the behaviour of electrons. After attending a colloquium in Munich in July where he and Schrödinger had disagreed publicly, Heisenberg wrote to Pauli, 'Schrödinger throws overboard everything which is "quantum-theoretical": namely, the photoelectric effect, the Franck collisions, the Stern–Gerlach effect, etc. It is not then difficult to establish a theory. However, it does not agree with experience' (Heisenberg to Pauli, 28 July 1926, Pauli, 1979, pp. 337–8). In a paper published in November, Heisenberg highlighted Schrödinger's failure to explain the quantised exchange of energy between atoms in resonance phenomena, through a continuous wave theory (Heisenberg, 1926c).

Aside from criticisms over the empirical adequacy of Schrödinger's interpretation of wave mechanics, Heisenberg emphasised that the ψ-function did not, in general, represent a wave-field in ordinary three-dimensional space, but rather a *system* in n-dimensional abstract configuration space where n represents the number of degrees of freedom of the system. Only in the simple case of the hydrogen atom with a single electron could one interpret the wave equation in three-dimensional space, neglecting the electron spin. In his letter to Pauli in June, Heisenberg remarked that according to Schrödinger, in the case of the hydrogen atom: 'One should imagine the rotating electron, whose charge is distributed over the entire space and which has an axis in a fourth and fifth dimension' (Heisenberg to Pauli, 8 June 1926, Pauli, 1979, p. 328 [item 136]). As he explained in September 1926, 'it is possible to establish a wave equation in the coordinate space of 3-f-dimensions for the problem of the motion of f-particles, which then completely replaces the quantum-mechanical problem mathematically' but the Schrödinger wave, Heisenberg insisted, cannot be interpreted as representing a 'real' matter wave in three-dimensional space (Heisenberg, 1926b, pp. 992–3). Indeed, as Heisenberg here suggests, it is more natural to interpret the ψ-function as describing a system of particles. Although in both the passage quoted above and in a paper on the many-body problem and resonance phenomena published in June, Heisenberg spoke of a system of particles, he admitted it cannot be described in terms of 'our usual space-time concepts' (Heisenberg, 1926a, pp. 412–13) and the concept of the individuality of a particle breaks down in quantum mechanics (Heisenberg, 1926b, p. 993). Even physicists like Einstein and Lorentz, who were greatly impressed by the discovery of wave mechanics, expressed similar concerns in their correspondence in 1926 about the extent to which we could speak of the 'reality' of the Schrödinger waves in configuration space (Lorentz to Schrödinger, 27 May 1926, Przibram, 1967, p. 44; Einstein to Lorentz, 1 May 1926 & Einstein to Ehrenfest, 28 August 1926, Mehra & Rechenberg, 2001, pp. 234–5).

Schrödinger was well aware of the difficulties posed by his theory, but was hopeful that such problems would in due course be overcome. In his second paper, he acknowledged that when treating 'the problem of several electrons' using the equation of wave mechanics, we must interpret the ψ-function 'in configuration space, and not in real space' (Schrödinger, 1928b, p. 60). The fact that the ψ-function is 'not a function of ordinary space and time as in ordinary wave problems', Schrödinger conceded, 'raises some difficulty in attaching a physical meaning to the wave function'. But these difficulties did not deter him from continuing the search for a wave theory of matter in three-dimensional space, which had thus far eluded physicists.

By mid-1926, Born would publish two important papers, in which he interpreted the ψ-function, representing the collision of an electron with an atom, not as a real wave, but as a probability wave (Born, 1926a, 1926b). As Abraham Pais put it, 'we owe to Born the beginning insight that ψ-itself, unlike the electromagnetic field, has no direct physical reality' (Pais, 1982, p. 1197). In late November, Born wrote to Einstein, explaining that the work carried out by Pauli and Jordan showing that for an atomic system of many electrons 'the probability field does not of course move in ordinary space', and to this extent 'Schrödinger's achievement reduces itself to something purely mathematical' (Born to Einstein, 30 November 1926, Pais, 1982, p. 1197). By the second half of 1927, Born's probabilistic interpretation of the ψ-function, now widely accepted by physicists, became a defining feature of what would later become known as the Copenhagen interpretation of quantum mechanics. We will return to this in the context of the wave–particle duality in the next chapter.

3.3 Drawing the battle lines: Heisenberg and Schrödinger on understanding in physics

The Heisenberg–Schrödinger clash can be understood in terms of the different attitudes to the physical meaning of the ψ-function. At a more fundamental level, their debate can also be interpreted as a disagreement as to what constitutes a physical interpretation of quantum mechanics (Chevalley, 1991b). By exploring this theme we can arrive at a deeper insight into Heisenberg's philosophical approach to the problem of interpretation. Schrödinger, whose philosophical thought has been the subject of much recent analysis (Bitbol & Darrigol, 1992; Götsch, 1992; Bitbol, 1996), took the view that in order to 'understand' a physical theory, such as quantum mechanics, one must be able to give a description in space and time – essentially a Kantian argument. Heisenberg, together with Bohr, felt it was necessary to adopt an altogether different attitude to the problem of what it meant to 'understand' nature.

The key philosophical issue that divided Heisenberg and Schrödinger in 1926 was what was meant by 'understanding' (*Verstehen*) in the context of quantum mechanics. In his second paper, Schrödinger launched a philosophical attack on Heisenberg, Born and Jordan for abandoning the possibility of describing the motion of electrons in space and time. Schrödinger, for his part, insisted that giving up all hope of a space-time description marked a complete renunciation of the task of 'understanding' the physical process within the atom. In his second paper on wave mechanics, Schrödinger had stressed that rejecting the concept of the electron orbit in quantum theory did not, as some physicists had maintained, spell the end of all space-time description:

> All these assertions systematically contribute to the relinquishing of the ideas of "position of the electron" and "path of the electron". If these are not given up, contradictions remain. This contradiction is so strongly felt that that it has been doubted whether the phenomena in the atom can be described in the space-time form of thought at all. From the philosophical standpoint, I would consider such a definitive decision of this sort to be equivalent to complete surrender. For we cannot really change our forms of thought, and what we cannot *understand* within them, we cannot *understand* at all. There *are* such things – but I do not believe that the structure of the atom is one of them.
>
> *(Schrödinger, 1928d, pp. 26–7, emphasis added)*

Schrödinger's appeal to the necessity of space and time as the 'forms of thought' through which we apprehend the world suggests a Kantian influence. This position would place him at odds with Bohr and Heisenberg. Writing to Wilhelm Wein in August 1926, after returning from his visit to Copenhagen, Schrödinger reiterated his stance on the necessity of a visualisable space-time representation of the moving electron in quantum mechanics: 'Bohr's standpoint', also shared by Heisenberg, 'that a space-time description is impossible, I reject *a limine*'. If the objects of empirical knowledge 'cannot be fitted into space and time', argued Schrödinger, then physics 'fails in its whole aim and one does not know what purpose it really serves' (Schrödinger to Wein, 25 August 1926, Moore, 1989, p. 226). The philosophical attitude underpinning the work of Bohr and Heisenberg was, in Schrödinger's view, a distortion of the true aim of physics.

The dispute between Schrödinger and Heisenberg centred on the necessity of representing the atom in the classical space-time framework. They disagreed about what would constitute an *understanding* of nature in the context of quantum mechanics. After his discussions with Bohr in Copenhagen, Schrödinger wrote to Wein complaining about the philosophical viewpoint he had encountered there: The 'approach to atomic problems' adopted by Bohr and Heisenberg 'is really remarkable'. They are 'completely convinced that any *understanding* in the usual sense of the word is impossible. Therefore the conversation is almost immediately driven into *philosophical* questions' (Schrödinger to Wein, 21 October 1926, Moore, 1989, p. 228, emphasis added). Heisenberg later recalled that while Bohr had for some time argued that it might yet be possible to 'understand' the atom, he stressed that 'in the process we may have to learn what the word "understanding" really means' (Heisenberg, 1971, p. 41). In his interview with Kuhn he explained:

> We had seen that we could, possibly understand nature, but at the same time we had to learn that the word "understand" means something different from what we had believed it would in the earlier times. So it is a whole change of attitude about what we can and cannot do. Understanding suddenly just means predicting experiments. "Understanding" does not mean something like in the old classical physics. That change took place already in the first stage of quantum mechanics, I would say in '24 already. We didn't want to go back to the old line, and that was the disappointment with Schrödinger ... Therefore I was so upset about the Schrödinger development in spite of its enormous success.
>
> *(AHQP, 22 February 1963, p. 30)*

Heisenberg's view that 'understanding suddenly just means predicting experiments' suggests that he had by this time resorted to an instrumentalist view of the theory. The task now seemed to be simply to 'save the phenomena'. Importantly, this did not reflect some prior commitment to positivism on Heisenberg's part, but a pragmatic approach that had been forced upon the physicists by the experimental phenomena. As he would explain in his lecture in September 1926, 'the program of quantum mechanics' is 'to establish simple relations between experimentally determined quantities' (Heisenberg, 1926b, p. 990). For Schrödinger, the interpretation of quantum mechanics took as its starting point the Kantian position that the forms of space and time are the conditions for the possibility of understanding. For Heisenberg, such a standpoint was no longer tenable. In his view, the classical ideal of understanding could not be allowed to dictate the form of interpretation. Rather, progress in physics would itself determine what is meant by 'understanding'. As he put it later: 'The exact sciences start from the assumption that in the end it will always be possible to understand nature, even in every new field of experience, but that we may make no *a priori* assumptions about the meaning of the word "understand"' (Heisenberg, 1958b, p. 28).[3] Put simply, Schrödinger was committed to the view that in physics, one *must* represent processes in space and time. It was this view that drove him to search for a wave theory of matter. Heisenberg, on the other hand, saw the renunciation of the classical ideal of understanding, and with it the representation in space and time, as *forced upon us* by quantum mechanics. Expressed differently, for Schrödinger, it was the *nature of understanding* that guided his approach to the interpretation problem, whereas for Heisenberg, it was only through the development of physics that one would learn what it means to *understand nature*.

Here it is critical to realise that for Heisenberg, it was the existence of *classical phenomena*, such as the observation of the tracks left by moving particles in the Wilson cloud chamber, which posed the biggest difficulty for the interpretation of quantum mechanics. Neither matrix mechanics nor wave mechanics could as yet account for this phenomenon. To this extent, the motion of *free* electrons, which appeared to move in definite trajectories, eluded quantum-mechanical treatment. Thus, only when this phenomenon, and for that matter all motion in space and time, could be explained 'quantum-mechanically' could one claim to have 'understood' the theory. This is precisely what Heisenberg was able to achieve in his March 1927 paper, in which he formulated for the first time the uncertainty principle (Heisenberg, 1927b).[4] Employing Born's statistical interpretation and the Dirac–Jordan transformation theory, Heisenberg was able to account not only for the quantum jumps between stationary states and the scattering of electrons in atomic collisions, but the 'motion' of free particles. In section three of his paper, Heisenberg explained that it was now possible to understand 'the transition from

---

[3] This view is made explicit in Heisenberg's 1932 lecture 'On the History of the Physical Interpretation of Nature', where he argued: 'Every great discovery – and this can be seen especially in modern physics – moderates the pretensions of scientists to an understanding of the universe in the original sense' (Heisenberg, 1952b, p. 34). In Chapter 2 of his work on the interpretation of quantum theory, James Cushing attempts to distinguish three levels at which scientific theories function – empirical adequacy, formal explanation and understanding (Cushing, 1994, pp. 9–23).

[4] Translated into English as 'The Physical Content of Quantum Kinematics and Mechanics' by Wheeler and Zurek (Heisenberg, 1983). I will refer to the English translation by Wheeler and Zurek in this chapter unless otherwise indicated.

micro- to macro-mechanics' (Heisenberg, 1983, pp. 72–6) and showed 'how the transition, discussed above, to classical theory is formulated mathematically for a simple mechanical system, the force-free motion of a particle' in one dimension (Heisenberg, 1983, p. 74). As is now well known, Heisenberg was able to accomplish this by interpreting the Schrödinger wave, not as a 'real' wave propagating through space, but rather as a probability wave in configuration space. In this way, the electron's motion in the Wilson cloud chamber was explained, not as a continuous trajectory through space, but as a series of discontinuous interactions with the molecules of water vapour in the cloud chamber.[5]

The 1927 paper thus marks the culmination of Heisenberg's program of extending the scope of matrix mechanics. Rather than attempt to interpret quantum mechanics by treating the electron classically as a moving point-charge or a matter wave, Heisenberg saw the task before him to reinterpret the seemingly *classical motion* of a particle quantum-mechanically, by bringing it within the mathematical framework of both matrix mechanics and wave mechanics. In his 1927 paper Heisenberg was able to show that the motion of a free particle would 'take place approximately as predicted classically', but importantly 'without violating the laws of quantum theory' (Heisenberg, 1983, p. 76). With this, Heisenberg felt that it was now possible to 'intuitively understand' the theory. As he put it in a letter to Pauli on 23 February, 'if one believes, as I do, that one *intuitively understands* the physical laws if one can say instinctively what comes out in every case, then such considerations may help a bit further; in any case they have eased my conscience' (Heisenberg to Pauli, 23 February 1927, Pauli, 1979, p. 381 [item 154], emphasis added). In the opening paragraph to his paper, he again presented his own definition of 'intuitive understanding':

> We believe that we *understand* the physical content of a theory *intuitively* when we can see its qualitative experimental consequences in all simple cases and when at the same time we have checked that the application of the theory never contains inner contradictions. For example we believe that we understand the physical content of Einstein's concept of a closed 3-dimensional space because we can visualize consistently the experimental consequences of this concept. Of course these consequences contradict our everyday physical concepts of space and time.
> *(Heisenberg, 1983, p. 62, emphasis added)*

The above passage is highly revealing for it makes it clear that the critical issue in the interpretation of quantum theory was what it means to 'understand' a physical theory. Significantly, Heisenberg drew a direct parallel with the

---

[5] In arriving at this conclusion Heisenberg later claimed to have been inspired by Einstein's view that 'it is the theory that decides what we can observe', as discussed in Chapter 2.

concept of 'understanding' that physicists had been forced to accept as a result of Einstein's theory of relativity. The meaning of the word 'understanding' in the context of the special and general theory of relativity had been the subject of discussions between Heisenberg and Pauli in the early 1920s (Heisenberg, 1971, pp. 29–33). But for Bohr, this shift to a formalistic definition of 'understanding' in physics moved in the wrong direction. As Heisenberg later recalled, Bohr 'was not too much impressed by mathematical consistency or non consistency. He would simply say, "Well, have I *understood* how the thing works?" And he would say to understand means to use the two pictures, waves and particles, and to play back and forth between the two pictures' (AHQP, 5 July 1963, p. 14, emphasis added). Bohr's view of wave–particle duality and the influence he exerted on Heisenberg will be considered in more detail in the next chapter. However, more remains to be said on the historical context of Heisenberg's redefinition of *Anschaulichkeit* in the 1927 paper.

3.4 Redefining *Anschaulichkeit*

Further light can be shed on Heisenberg's critical response to Schrödinger, by shifting our attention from 'understanding' (*Verstehen*) to 'visualisability' (*Anschaulichkeit*). The German term *Anschaulichkeit*, which appears frequently in Schrödinger's writings, does not, as Henk de Regt explains, 'only mean "visualisability" but also "intelligibility"' and is to this extent closely connected with the notion of 'understanding' in German. For this reason, de Regt points out, 'the notion of *Anschaulichkeit* played a crucial role in the genesis of quantum mechanics' (de Regt, 1997, pp. 461–2). In a footnote to his second paper, Schrödinger explained that he had been 'discouraged, not to say repelled' by the 'want of *Anschaulichkeit*' in the matrix theory developed by Heisenberg (Schrödinger, 1928b, p. 46). As Linda Wessels has explained, Schrödinger saw himself as working in the tradition of physicists such as Herz and Boltzmann, who had emphasised the necessity of visualisability for understanding (Wessels, 1993, pp. 270–1). By contrast, Heisenberg, we will recall, had stressed that quantum mechanics was not 'directly amenable to a geometrically *anschaulich* interpretation, since the motion of electrons cannot be described in terms of the familiar concepts of space and time' (Born, Heisenberg & Jordan, 1967, p. 322). To give a geometrically *anschaulich* interpretation of quantum mechanics evidently meant to give a description of the electron's motion in space and time.

However, in his 1927 paper in which he formulated the uncertainty relations for the first time, Heisenberg adopted a somewhat different stance with regard to *Anschaulichkeit*. Having earlier insisted that quantum mechanics was not amenable to a visualisable interpretation, he now claimed that the matrix representation of the position coordinate of an electron in quantum theory 'is perhaps no more abstract and no more unintuitive [*unanschauliche*]' than the representation of the electric field by an anti-symmetric tensor in the general theory of relativity (Heisenberg, 1983, p. 82). Significantly in this context, the title of the paper was '*Über den anschaulichen Inhalt der quantentheoretische Kinematik und Mechanik*', which translates roughly as 'On the Intuitive Content of Quantum-Theoretical Kinematics and Mechanics'. As a number of scholars have argued, the shift in Heisenberg's attitude turns on his subtle redefinition of the German word *anschaulich* (Miller, 1978, 1982; Hendry, 1984, p. 125; Miller, 1986, pp. 127–83; Beller, 1992b, p. 285; Lacki, 2002, pp. 248–451). In the concluding section of the paper, Heisenberg makes this clear: 'As we can think through qualitatively the experimental consequences of the theory in all simple cases, we will no longer have to look at quantum mechanics as *unanschaulich* or abstract' (Heisenberg, 1983, p. 82). The concept of *Anschaulichkeit* now took on a different meaning more in tune with an empiricist viewpoint.

While scholars have rightly pointed out that the redefinition of *Anschaulichkeit* in the 1927 paper must be viewed in the context of Heisenberg's debate with Schrödinger, they have ignored the wider historical and philosophical context in which the debate over 'visualisability' was situated. Mara Beller argues that the redefinition of visualisation in the 1927 paper was 'not an argument for *Anschaulichkeit* as Heisenberg's contemporaries would have understood it' Beller, 1999, p. 115). Taking much the same view, Paul Forman argues that Heisenberg's redefinition of the term was nothing more than a rhetorical device which served to appease his audience who were hostile to the idea of a non-intuitive theory of the micro-world:

> Heisenberg sought to remove the stigma of *Unanschaulichkeit* by redefining the "intuitive" quality so as it make it predictable of his irremediably unpictorial quantum mechanics. This redefinition equated "intuitive" to "satisfactory" in a strictly positivist sense … Perhaps the best index of the wantonness of this solution is that in their colleagues' eyes Heisenberg and Born (who joined in this redefinition) had simply inverted the ordinary, accepted signification of the word.
>
> *(Forman, 1984, pp. 339–40)*

Forman is quite right to point out that many physicists like Arnold Sommerfeld continued to refer to quantum mechanics as *unanschauliche* even after the publication of Heisenberg's paper (Sommerfeld, 1930). Yet, the meaning of *Anschaulichkeit* that we find in Heisenberg's 1927 paper, or at least something very close to it, can also be found in an empiricist strand of thought that emerged in the German-speaking world after the development of non-Euclidean geometry in the second half of the nineteenth century and the discovery of Einstein's general theory of relativity. To this extent, Heisenberg did not employ the term wantonly, but he used it in a sense that was becoming widely accepted by many of his contemporaries with empiricist philosophical leanings. This becomes clearer if we remember that originally the term *Anschaulichkeit* derived its meaning from Kantian philosophy, in particular the transcendental view of space and time as conditions of the possibility of experience. As Arthur Miller rightly points out, 'Owing to the complexity of Kant's philosophy the term *Anschauung* became virtually untranslatable into English, and in any language it had to be understood in the context in which Kant's philosophy was discussed' (Miller, 1986, p. 127). To this extent, Heisenberg's redefinition of *Anschaulichkeit* can be viewed as part of the widespread reaction to Kantian philosophy in the first decades of the twentieth century.

Catherine Chevalley has discussed the shifting conception of *Anschaulichkeit* in nineteenth century German philosophy, though she has not connected this with Heisenberg's redefinition of *Anschaulichkeit* in his 1927 paper (Chevalley, 1994, pp. 33–55). The emergence of non-Euclidean geometry in the second half of the nineteenth century had already prompted a number of thinkers to propose that Kant's notion of *Anschaulichkeit* would need to be expanded, redefined or altogether discarded. In an important paper published in 1876 entitled 'The Origin and Meaning of Geometrical Axioms', Hermann von Helmholtz argued that one must modify the Kantian concept of *Anschauung*. He described the human capacity to 'visualise' non-Euclidean geometry in the following terms: 'we have no word but *intuition* to mark this; but it is *knowledge empirically gained* by the aggregation and reinforcement of similar recurrent impressions in memory, and not a transcendental form given before experience' (Helmholtz, 1876, p. 320, emphasis added). In another article published just two years later, Helmholtz defended this empiricist definition of 'visualizability': 'I am of the opinion that this definition contains stricter and more definite requirements for the possibility of imagination than any previous one' (Helmholtz, 1878, p. 215). In the German version of this article '*Über den Usprung und Sinn der geomterische Sätze*' which appeared in volume two of Helmholtz' *Wissenschaftliche Abhandlungen*, the word 'imagination' is translated as '*Anschauung*' signifying a direct connection with the Kantian philosophy (Helmholtz, 1883, p. 644).[6] A striking parallel exists between this definition and the one that appears in Heisenberg's 1927 paper. Indeed, in his paper Heisenberg explicitly drew an analogy between his own conception of 'intuitive understanding', and that found in the use of non-Euclidean geometry in the special theory of relativity.

---

[6] Robert di Salle has made a thorough examination of Helmholtz' redefinition of the Kantian notion of *Anschaulichkeit* (di Salle, 1993).

We should not overlook the fact that Helmholtz' writings exerted a considerable influence on the German-speaking world around this time. In 1921, Paul Hertz and Moritz Schlick published a selection of Helmholtz' papers and lectures on the philosophical foundations of the sciences (Helmholtz, 1977). His empiricist notion of visualisation, as we shall see shortly, also appears to have left its mark on Einstein around this time. Helmholtz' critique of Kant's concept of pure intuition of space was taken up in the early twentieth century by German philosophers steeped in the neo-Kantian tradition, notably Cohen, Riehl, and Cassirer, all of whom attempted to defend or modify the Kantian concept of *Anschauung* (Coffa, 1991, pp. 41–61). In his discussion of non-Euclidean physical space in 1927, Hermann Weyl explicitly referred to Helmholtz' critique of Kant's notion of *Anschaulichkeit*. Weyl pointed out that Helmholtz had taken issue with the Kantian view that 'non-Euclidean geometry is devoid of intuitivity (*Anschaulichkeit*)', in proposing a new 'definition of intuitivity' (Weyl, 1949, p. 133). In his widely read 1921 paper delivered to the Prussian Academy of Sciences, entitled 'Geometry and Experience', Einstein examined the question of whether it was really possible to 'visualise' a finite, yet unbounded space in three dimensions:

> A geometrical-physical theory [of this kind] as such is incapable of being directly pictured, being merely a system of concepts. But these concepts serve the purpose of bringing the multiplicity of real or imaginary sensory experiences into connection in the mind. To "visualise" a theory, or to bring it home to one's mind, therefore means to give a representation to that abundance of experiences for which the theory supplies the schematic arrangement.
>
> *(Einstein, 1996, pp. 115–16)*

The definition of 'visualisation' presented here is similar to the one we find in the introduction to Heisenberg's 1927 paper. In presenting his idea of 'intuitive understanding', Heisenberg explicitly invokes Einstein's concept of closed three-dimensional space. The passage quoted above strongly indicates that Heisenberg's view bears the critical influence of Einstein's own redefinition of *Anschaulichkeit*. Einstein's views would exert a considerable influence on logical empiricism throughout the 1920s and 1930s. In 1922, Hans Reichenbach explained that the 'problem of intuition' had aroused much philosophical discussion in the context of Einstein's theory of relativity (Reichenbach, 1996, pp. 280–3). But it was already becoming apparent that quantum theory would demand further reflection on this question. Defending an empiricist view against the neo-Kantian standpoint, Reichenbach commented on the implications of the newly emerging quantum theory: 'If physics should proceed, under the influence of quantum theory, to conceive of space as a discrete manifold (a matter that is undecidable at the moment)', then the 'concept of pure intuition [*Anschauung*] would require further extension' Reichenbach, 1996, p. 275).

In a letter to Bohr in December 1924, Pauli had criticised those 'weak' physicists like Kramers, who felt the need to visualise the electron as moving in a classical orbit in the atom. The demand for a classical visualisation, Pauli stressed, 'should still never count in physics for the retention of a certain set of concepts. When the system of concepts is once clarified, *then there will be a new visualisation*' (Pauli to Bohr, 12 December 1924, Pauli, 1979, pp. 186–9 [item 74].) Pauli's brief remark on the close theoretical connection between 'visualisation' and the 'system of concepts' bears a striking resemblance to the view expressed by Einstein in the passage quoted above. In 1928, Philip Franck, one of the leading figures of the Vienna Circle, published a paper focusing specifically on the meaning of *Anschaulichkeit* in modern physics. Throwing his support behind Heisenberg's redefinition of the term, Franck saw this as further confirmation of the positivist direction of modern physics, spelling the death knell for metaphysical materialism and Kantian idealism (Frank, 1928). The redefinition of *Anschaulichkeit* would play a major part in the evolving physical discourse of the 1920s, to which both philosophers and physicists would contribute.

The connection between a system of concepts and visualisation, found in both Einstein and Pauli, is also evident in the introduction to the paper Born and Heisenberg presented for the Solvay conference in October 1927 – 'The new *system of concepts* gives at the same time the *intuitive* content of the new theory'. Defining an 'intuitive theory' as one that is 'in itself without contradiction' and 'permits one to predict without ambiguity the results of all imaginable experiences in its domain', Born and Heisenberg now argued that quantum mechanics must be regarded as 'an intuitive and complete theory of micromechanics' (Born & Heisenberg, 1928, p. 144). Here one discerns an empiricist line of thought running from Helmholtz through Einstein, to Heisenberg.

Bohr, however, did not share Heisenberg's reinterpretation of *Anschaulichkeit*. While the two physicists had reached agreement on the interpretation of quantum-mechanical formalism by the spring of 1927, there remained differences of opinion on the visualisability of the new theory. Writing to Pauli in May 1927, Heisenberg reported that in his discussions in Copenhagen, 'there are at present between Bohr and I essential differences of opinion over the word *anschaulich*' (Heisenberg to Pauli, 16 May 1927, Pauli, 1979, pp. 394–6 [item 163]). As Beller puts it, 'For Bohr, with his direct, down-to-earth physical intuition, Heisenberg's sophisticated redefinition of *Anschaulichkeit* must have seemed an illegitimate move. This was, in fact, one of the central issues in the clash between Heisenberg and Bohr over the uncertainty paper' (Beller, 1992b, p. 285). In a similar vein, Hendry points out that for Bohr, 'the requirements of *Anschaulichkeit* were much stricter, and could be satisfied only by a fully consistent classical visualisation' (Hendry, 1984, p. 124). Bohr's unpublished manuscripts written in the summer of 1927 make it clear that he approached the problem of visualisation differently from Heisenberg (Murdoch, 1987, p. 78). In the 1930s, Bohr returned to 'the often discussed question' of whether or not the mathematical formalism of quantum mechanics 'can be regarded as an extension of our power of visualization', but as before he rejected such a possibility (Bohr, 1937, p. 292). Instead, Bohr insisted that by abandoning a description in space and time, quantum mechanics demands a 'radical renunciation of the usual claims of so-called visualization' (Bohr, 1937, p. 294).

After the 1927 paper, Heisenberg did on several occasions revert to the more conventional use of the term to refer to classical visualisation in space and time. This is evident in a short paper he published in April 1927, where Heisenberg declared that the 'motions of electrons are *not intuitively* describable' [*nicht anschaulich beschreibbar*] in the sense that they cannot be described in the familiar concepts of space and time (Heisenberg, 1927b, p. 83). Here, we should take into account that Heisenberg was writing for a different audience – one not so familiar with the historical and philosophical context in which the term *Anschaulichkeit* had acquired new meaning. In writing the introduction to the uncertainty paper, Heisenberg was quite consciously drawing on the lessons of general relativity and non-Euclidean geometry. The revolution brought about by Einstein's theory of relativity, for Heisenberg, had not only transformed our basic concepts of the space and time, but had in the process altered the very meaning of 'intuitive understanding'. For Heisenberg, this new attitude had to be carried over in quantum mechanics.

3.5 Instrumentalism or realism? Heisenberg's notion of closed theories

Undoubtedly, there was a strongly positivist-empiricist element in Heisenberg's philosophy in the 1920s. This is not to say that Heisenberg accepted a completely anti-realist view of science. As we saw in the previous chapter, by October 1927 he did not feel it was possible to draw a sharp distinction between the unobservable and observable elements of a physical theory. He did however see it as the task of physics to provide a consistent mathematical description which 'saved the phenomena'. In defending the statistical interpretation of quantum mechanics Heisenberg concluded his 1927 paper on a typically positivistic note: the assumption that 'there still hides a "real" world in which causality holds', he argued, would 'seem to us, to say it explicitly, fruitless and meaningless. Physics ought to describe only the correlation of observations' (Heisenberg, 1983, p. 83). Passages such as these appear to reflect the influence of logical positivism in the late 1920s, though they may be read as Heisenberg's attempt to *defend* or *legitimise* the new theory of quantum mechanics against its detractors.

Yet, as Mara Beller has noted, there are several passages in Heisenberg's writings from this time that appear to support a realist view. Heisenberg's 'realist' voice comes out most strongly in the concluding section of the 1927 paper. Here Heisenberg argued that 'the velocity [of an electron] in the X-direction is "in reality" not a number, but the diagonal term of a matrix'. Here, he maintained that 'all quantum-theoretical quantities are "in reality" matrices' (Heisenberg, 1983, p. 82). In this context, Beller draws the conclusion that: 'No coherent philosophical choice between positivism and realism guided Heisenberg's efforts. A fascinating, ever changing mixture of realist intuition and positivist legitimation characterises Heisenberg's work leading to, and springing from, the [1925] reinterpretation paper' (Beller, 1999, p. 52). In his later writings, Beller argues, 'Heisenberg adopted Bohr's positivist approach for mathematically unsophisticated audiences, yet he employed elements of a realistic ontological interpretation when addressing his mathematically skilled colleagues'. To this extent, Beller argues that 'two contradictory approaches were present simultaneously' in Heisenberg's writings on the interpretation of quantum mechanics (Beller, 1999, p. 172).

How should we read the apparent contradiction between positivism and realism in Heisenberg's thought? Was he simply confused? Or did he just express himself carelessly at such points? In reading the 1927 paper we should bear in mind that Heisenberg's philosophical thinking was in a state of flux during this time (Beller, 1999, pp. 103–16). As we saw in the previous chapter, Heisenberg's commitment to the principle of observability, which was probably never very strong to begin with, became weaker after his discussions with Einstein in April 1926. However, the realist and positivist impulses in Heisenberg's thought need further explanation. In my view, we can actually make good sense of Heisenberg's viewpoint if we pay careful attention to precisely in what sense he declares himself to be a 'realist'. Here, I am not claiming to resolve all the contradictions which appear in Heisenberg's 1927 paper – as we shall see in Chapter 5, Heisenberg's views about the wave–particle duality and how we assign meaning to concepts like position and momentum continued to evolve during this time – but I feel we can understand Heisenberg's outlook better once we realise that the view which originates in this paper, but which comes out more clearly in later writings, is in many respects close to a view commonly referred to today as 'structural realism'.

In order to see this, it is necessary that we delve into Heisenberg's concept of a closed theory in physics, which has recently been the subject of some important work by Alisa Bokulich (2004, 2006) and Melanie Frappier (2004). Heisenberg's first detailed discussion of his notion of a closed theory is found in a paper he published for the 1948 *Dialectica* volume (Heisenberg, 1948). However, there is good evidence to suggest Heisenberg had begun to conceive of quantum mechanics as a 'closed theory' as early as 1927. Indeed, its axiomatic structure was, for Heisenberg, closely connected with its 'completeness', a point he makes in the conclusion to the paper with Born which was presented at the 1927 Solvay conference. In the concluding section of that paper, quantum mechanics is described as 'a complete theory, whose fundamental physical and mathematical hypotheses are no longer susceptible to modification' (Born & Heisenberg, 1928, p. 178). This, as we shall see, comes very close to Heisenberg's later definition of a 'closed theory'.

According to Heisenberg, his thoughts on the idea of closed theories in physics crystallised during his visit to Chicago in 1929, during which time he had the opportunity to engage in philosophical discussions with the American physicist Barton Hoag. In these discussions, Heisenberg was led to outline his notion of a closed theory: 'The most important criterion for a closed system is probably the presence of a precisely formulated and self-consistent set of axioms governing the concepts and logical relations of the system'. Here, Heisenberg distinguishes four 'closed realms' in physics: Newtonian mechanics, statistical thermodynamics, relativistic electrodynamics, and finally quantum mechanics. 'For each of these realms there is a precisely formulated system of concepts and axioms, whose propositions are strictly valid within the particular realm of experience they describe' (Heisenberg, 1971, pp. 97–8). The key point for Heisenberg was that it is not possible to 'modify' or 'improve' a closed theory like Newtonian mechanics through slight adjustments to the basic equations. Where the concepts of classical physics no longer grasp reality, one must devise an entirely new set of concepts and relations with a new axiomatic structure. But this in no way shows classical physics to be invalid, only that there are definite limits to its applicability.

In the account Heisenberg presents us of his conversation with Hoag, he endeavoured to distinguish his own viewpoint from the more pragmatic attitude to physics he encountered in the United States. Here, Heisenberg attempted to explain what he saw as the real virtue of 'closed theories', as opposed to mere phenomenological theories, which he took to form the starting point of virtually all work in theoretical physics. Without admitting the existence of these closed realms in physics, 'we would lose the most important truth criterion of physics, namely the ultimate simplicity of all physical laws' (Heisenberg, 1971, p. 99). Heisenberg freely admitted here to being strongly attracted to the idea that 'the simplicity and beauty of the mathematical schemes' in physics give us every indication that they represent part of the structure of the world itself, and not merely our knowledge of it (Heisenberg, 1971, pp. 68–9). To this extent, the extraordinary simplicity of the theories like Newtonian mechanics and quantum mechanics, when compared with the wide range of phenomena they account for, was for Heisenberg of decisive importance. As he explained:

> If as we must always do as a first step in theoretical physics, we combine the results of experiments and formulae and arrive at a phenomenological description of the processes involved, we gain the impression that we have invented the formulae ourselves. If however, we chance upon one of those very simple, wide relationships that must later be incorporated into the axiom system, then things look quite different. Then we are suddenly brought face to face with a relationship that has always existed, and that was quite obviously not invented by us or by anyone else. *Such relationships are probably the real content of our science.*
>
> *(Heisenberg, 1971, p. 99, emphasis added)*

In an earlier passage in *Physics and Beyond*, Heisenberg had emphasised this same point in relation to the discovery of quantum mechanics:

> If nature leads us to mathematical forms of great simplicity and beauty – by forms I am referring to coherent systems of hypotheses, axioms, etc. – to forms that no one has previously encountered, we cannot help thinking that they are "true", that *they reveal a genuine feature of nature*. It may be true that these forms also cover our subjective relationship to nature, that they reveal elements in our own thought economy. But the mere fact that we could never have arrived at these forms by ourselves, that they were revealed to us by nature, suggests strongly that *they must be part of reality itself, not just of our thought about reality.*
>
> *(Heisenberg, 1971, p. 68)*

Heisenberg reports that he expressed this view, or something close to it, in his discussions with Einstein in April 1926, during which he had attempted to explain why he placed so much faith in matrix mechanics. As Alisa Bokulich rightly points out, we should carefully distinguish between Heisenberg's 'positivistic sounding statements ... concerning how one should go about constructing new scientific theories' and 'his views regarding the status of the final product of that methodology' (Bokulich, 2006, p. 100). This is the key to understanding Heisenberg's realism. While the physicist may employ a 'pragmatic' or even 'positivistic' view in searching for the laws of physics, if one uncovers a 'closed axiomatic system' which is able to account for a wide range of phenomena, we are entitled to interpret the theory as capturing some structural aspect of reality itself. To this extent, in Heisenberg's view, Newtonian mechanics, no less than quantum mechanics, employs a 'system of definitions and axioms' which can be regarded 'as describing an eternal structure of nature' (Heisenberg, 1958d, p. 93).

Far from subscribing to the Machian view that in the end all theories in physics are simply 'thought-economical' instruments through which we order our experience, Heisenberg argues that fundamental theories in physics allow us to know something about the *mathematical form* or *structure* of the physical world, though they do not give us direct knowledge of the *nature* or *essence* of the entities of physics. The laws of electrodynamics and quantum mechanics reveal an important aspect of the truth about the world, but we cannot arrive at a definite answer to the question of what an electromagnetic field or an electron is. As Heisenberg would put it: 'The philosophic content of a science is only preserved if science is conscious of its limits … Only by leaving open the ultimate essence of a body, of matter, or energy, etc., can physics reach an understanding of the individual properties of the phenomena that designate these concepts' (Heisenberg, 1958c, pp. 180–1). This view is, in many respects, similar to the view of 'structural realism', much discussed in the recent work of John Worrall (1989), James Ladyman (1998) and Stathis Psillos (2001). Indeed, as Barry Gower points out, 'the idea of structural realism, broadly construed, was considered and adopted by a number of philosophers in the early decades of this [the twentieth] century' (Gower, 2000, p. 74). Here, Gower identifies anticipations of the modern viewpoint of structural realism in the writings of Cassirer, Schlick, Carnap and Russell, all of whom were contemporaries of Heisenberg.

Yet, unlike many defenders of structural realism today, Heisenberg never seems to have emphasised the unity of physics, or even of reality itself. Indeed, Heisenberg was more inclined to speak of different physical theories such as classical Newtonian mechanics and statistical thermodynamics as describing different 'regions of reality'. One of the underlying themes of the 1942 manuscript was that in describing reality we cannot really speak of a single, unified reality, which merely appears in different guises (Heisenberg, 1984e, p. 221). As he was to put it: 'By the expression "region of reality" we mean a collection of nomological connections' (Heisenberg, 1984e, p. 233). Without such a unified collection of laws, 'one could not legitimately speak of a "region"', nor could one demarcate this region from others 'in order to render possible a division of reality'.

This 'structuralist' reading of Heisenberg enables us to make better sense of some of his later writings. In *Physics and Philosophy*, Heisenberg argued that: 'The "thing-in-itself" is for the atomic physicist, if he uses this concept at all, finally a mathematical structure; but this structure is – contrary to Kant – indirectly deduced from experience' (Heisenberg, 1958d, p. 83). It is important to realise that for Heisenberg the ψ-function does not represent an 'objective reality' in space and time, nor does it represent a *Ding an sich* in the sense of some noumenal reality existing 'behind the phenomena'.[7] Rather, it must be understood as a *mathematical form* revealing the structure of an underlying reality revealed to us through experience. In his later years, Heisenberg saw himself as following in the 'Platonic' tradition, according to which the ultimate task of physics is to find the underlying mathematical forms in nature. This line of thought finds its clearest expression in his philosophical writings on the non-linear field theory of elementary particles in the 1950s and 1960s where he suggests that modern physics is closer in spirit to Platonic idealism than the atomism of Democritus (Heisenberg, 1971, pp. 237–47). In a letter to his sister-in-law, Edith Kuby, written in 1958, Heisenberg expressed his awe at the 'incredible degree of simplicity' of the laws of unified field theory: 'Not even Plato could have believed them to be so beautiful. For these interrelationships cannot be invented; they have been there since the creation of the world' (E. Heisenberg, 1984, p. 144). This view of the relationship of mathematics to reality stands in contrast to that of Bohr, who by and large looked upon mathematical formalisms in physics as useful tools in ordering our experience of the world (Murdoch, 1987, pp. 216–17). These different attitudes were at the heart of the disagreement between Bohr and Heisenberg that surfaced in the late 1920s regarding the interpretation of quantum mechanics.

The debate between Heisenberg and Schrödinger in 1926 has long been viewed as one of the defining episodes in the history of quantum mechanics. At the heart of their different approaches to the interpretation problem were fundamentally different *philosophical* standpoints concerning the meaning of 'understanding' in physics. Whereas Heisenberg held that a classical wave theory of the electron's motion in space and time was impossible, Schrödinger took as his starting point, the view that visualisation in space and time was a necessary condition for 'understanding' in physics. According to Heisenberg, quantum mechanics demands we abandon all hope of a classical understanding, and instead we must learn from physics itself what the word 'understanding' really means. In 1927, Heisenberg argued that one could 'intuitively understand' quantum mechanics, firstly because it was now possible to 'see its qualitative experimental consequences in all simple cases' and secondly because 'the application of the theory never contains inner contradictions'. This definition of 'intuitive understanding' reflects the influence of an empiricist strand of thought which finds expression in the earlier works of Helmholtz and Einstein in their

---

[7] Heisenberg stressed this point in discussions with the logical positivists following his Vienna lecture in 1930 (Diskussion über Kausalität und Quantenmechanik, 1931, pp. 186–7).

writings on the visualisation of non-Euclidean geometry. Heisenberg's redefinition of the term *Anschaulichkeit* (or visualisability) bears a striking resemblance to the way in which the term, traditionally associated with Kant's philosophy, was reinterpreted by Helmholtz and Einstein. This is significant, for it situates Heisenberg in the context of an empiricist reaction to Kant's philosophy, which had gathered considerable momentum in the German-speaking world in the 1920s. Yet, as we have indicated, Heisenberg was no instrumentalist in any straightforward sense of the term. His position was in some respects closer to one we would label today as structural realism.

We should not infer from all this that Heisenberg was uninterested in the conceptual framework for describing the 'kind of reality' of electrons and atoms. As we shall see in Chapter 4, Heisenberg's early attempts to interpret quantum mechanics revolved around his understanding of the limited applicability of classical concepts like particle and wave in the quantum world. Through his discussions with Bohr in Copenhagen in 1926–7, Heisenberg came to realise that it was no longer clear to what extent we could really speak of the electron as a 'particle' or a 'wave'. An electron seemed to behave as a particle in some certain experiments, but in others, it appeared to exhibit interference and diffraction effects more characteristic of wave phenomena. This paradox, for which physicists had no definitive answer in 1926, would become the subject of serious discussions in early 1927 between Bohr and Heisenberg. Bohr's attempt to resolve the wave–particle dilemma in quantum theory has been the subject of much scholarly attention, yet surprisingly little has been devoted to Heisenberg's own quite distinct notion of wave–particle *equivalence*. As we shall see in the next chapter, this way of conceptualising the duality in quantum mechanics only became clear to Heisenberg after the appearance of the Jordan–Klein–Wigner papers on quantum electrodynamics in 1928.

PART II

The Heisenberg–Bohr dialogue

4

The wave–particle duality

As the previous two chapters have shown, Heisenberg's early inclination towards positivism must be understood in a qualified sense. Heisenberg was neither an operationalist nor an instrumentalist in any straightforward sense of these terms. But in order to fully understand the development of his thought during this critical period between 1926 and 1929, we must look more closely at his dialogue with Bohr, and the central themes which emerged during this time. These were, in my view, wave–particle duality, indeterminacy and the limitations of classical concepts, and the concept of complementarity, which will be explored in the next three chapters. This chapter deals specifically with the first of these – the wave–particle duality.

Schrödinger's visit to Copenhagen in September 1926 provided the catalyst for the intense discussions between Bohr and Heisenberg on the interpretation of quantum mechanics. A central theme of these discussions was the extent to which an electron could be described as both a wave and a particle. This was an issue that clearly had dominated Bohr's thinking in 1926–7. As Heisenberg later recalled, Bohr was convinced that wave–particle duality was 'the central point of the whole problem', and that one had 'to start from that side of the problem in order to understand it' (AHQP, 25 February 1963, p. 18). Bohr's eventual resolution of the wave–particle dilemma of light and matter is generally acknowledged as one of the central doctrines of the Copenhagen interpretation of quantum mechanics. As Barbara Stepansky puts it: 'The wave–particle duality has been accepted as a part of quantum mechanics that is set within the context of complementarity and referred to as the Copenhagen Interpretation' (Stepansky, 1997, p. 385). While Bohr's view has been the subject of much detailed historical and philosophical scholarship (Murdoch, 1987; Beller, 1992a,b; Held, 1994), little attention has been devoted to alternative views of the wave–particle duality, particularly amongst physicists who defended the 'orthodox' interpretation of quantum mechanics. This is largely because much of the literature on the history of wave–particle duality assumes that after 1927, Bohr's idea of wave–particle complementarity was widely accepted by his colleagues, in particular Heisenberg. The historical studies of Mehra and Rechenberg (2000, pp. 271–2), Jammer (1974, p. 69) and Stepansky (1997, p. 384) all rehearse the standard view that after an initial period of disagreement in 1927, Heisenberg accepted Bohr's notion of wave–particle complementarity.

However, this view has recently been called into question by Mara Beller, who has argued that 'a genuine unanimity of opinion' between Bohr and Heisenberg 'never occurred' (Beller, 1999, p. 226). Beller's work is important in recognising that there were key differences between Heisenberg and Bohr on the wave–particle duality, but precisely what Heisenberg's views on this question were remains the source of much confusion and ambiguity. While scholars have devoted considerable attention to Bohr's notion of complementarity, much less is known about Heisenberg's view of the wave–particle duality in quantum mechanics, for which he coined the term wave–particle *equivalence*. This is unfortunate because Heisenberg's view represents a unique and original perspective on the wave–particle duality of matter. This chapter examines the development and meaning of Heisenberg's notion of wave–particle equivalence between 1926 and 1928 before analysing the way in which Heisenberg's own view differed from Bohr's understanding of the wave–particle duality.

Heisenberg's concept of equivalence brings together two distinct approaches. The first of these is the statistical interpretation of the ψ-function originating in Born's papers on collisions in mid-1926, and subsequently developed by Pauli and Heisenberg (Pauli, 1927). According to this standard interpretation, the electron is a *particle* (though without a well-defined position and momentum). The second, lesser known, interpretation is that of quantised matter waves offered by Jordan, Wigner and Klein in 1927–8, whereby the electron is a quantised *wave-field* (though without a well-defined wave amplitude). Both approaches turned out to be mathematically equivalent and lead to the same results, convincing Heisenberg that the wave and particle representations of the electron were simply different ways of describing the same thing. In bringing a new perspective to Heisenberg's interpretation of the wave–particle duality, we can better understand the way in which Heisenberg's philosophy of quantum mechanics diverged from that of Bohr.

Though Heisenberg often gave the impression that he was in general agreement with Bohr, their interpretations actually differed quite markedly. For Heisenberg, the wave–particle duality rests on the different, but empirically equivalent, transformations of the quantum-mechanical *equation of motion*, whereas for Bohr, wave–particle duality arises from the two ways in which the electron manifests itself in different *experimental arrangements*. The significance of this divergence can be seen in the different ways in which Bohr and Heisenberg understood the task of finding a coherent interpretation of quantum mechanics. Heisenberg's approach was focused on the meaning of the abstract mathematical formalism. Bohr, on the other hand, was more concerned with trying to understand the apparently contradictory experimental phenomena, which seemed to require both wave and particle interpretations.

4.1 The physical reality of the electron: wave or particle?

In the months following Schrödinger's papers on wave mechanics in 1926, the wave–particle duality of matter emerged as the central problem for these attempting to make sense of the new theory. However, not all physicists were convinced of the need to abandon the idea that an electron is a point-charge moving, whose trajectory through space can be followed. The French physicist Louis de Broglie was critical of the view that 'we can no longer speak of the material particle describing a trajectory' in the atom (de Broglie, 1928a, p. 64). The task for de Broglie was to develop a new formulation of wave mechanics by finding a continuous wave equation in three-dimensional space which 'can be thought of as directing the motion of the particle' along a well-defined trajectory. The wave-field could then be thought of as 'a guiding wave' (de Broglie, 1928b, p. 135). In the discussion that took place at the Solvay conference in October 1927, Hendrick Lorentz stated that quantum theory should be based on the concept of a point-particle: 'For me, an electron is a corpuscle that at a given instant is located at a definite point of space, and if I have the idea that at the following moment this corpuscle is located elsewhere, I should think of its trajectory, which is a line in space'. Lorentz conceded that it was 'evidently possible that this theory would be very difficult to develop', but insisted that 'it does not appear to me *a priori* impossible' (*Électrons et Photons*, 1928, p. 248). By the end of 1927, however, the views of Bohr, Heisenberg, Pauli, Born, Dirac and Jordan had prevailed, effectively ending the efforts of physicists like de Broglie to find a semi-classical interpretation.[1] Yet, it is not at all clear that a

---

[1] In the 1950s the American physicist David Bohm developed a new interpretation of quantum mechanics, along the lines proposed by de Broglie in 1927, in which he reintroduced the concept of the particle trajectory into quantum mechanics. Encouraged by Bohm's work which appeared to overcome many of the objections initially raised against the theory at the 1927 Solvay conference, de Broglie returned to his original program of developing a pilot wave of the electron in the early 1950s. While this interpretation of quantum mechanics has received only limited attention, there has been a revival of interest in Bohm's theory in the last two decades evidenced in the works of Peter Holland (1993) and James Cushing (1994).

consensus had emerged among the leading physicists of the day on how to understand the wave–particle duality in quantum mechanics.

The problem of wave–particle duality had been one of the key issues animating debates on the interpretation of quantum mechanics in the discussions that took place during Schrödinger's visit to Copenhagen in September 1926. Writing to Fowler on October 26, 1926, Bohr remarked that his discussions with Schrödinger had 'gradually centred themselves on the problems of physical reality' posed by quantum mechanics (Bohr, 1985, p. 14). Again, in a letter to Kronig written just days later, Bohr highlighted the importance of Schrödinger's visit, suggesting that it had given 'rise to much discussion regarding the *physical reality*' of the electron in quantum theory (AHQP, Bohr to Kronig, 28 October 1926). Shortly after these exchanges, Heisenberg presented perhaps the clearest expression of what he saw as the problem of reality posed by the theory of quantum mechanics. In a lecture delivered in Düsseldorf on 23 September 1926, he declared:

> Electrons and/or atoms do not possess that degree of immediate reality that pertains to objects of daily experience. It is the very subject matter of atomic physics and, with it, also of quantum mechanics, to investigate the kind of physical reality applicable to electrons and atoms … [Although until now, electrons have been conceptualised as particles that move in well-defined orbits in the atom,] indications are that electrons have a similar degree of reality as do light quanta … Here we merely wanted to point out that the investigation of this typically discontinuous element and that "kind of reality" is the real problem of atomic physics and therefore the subject matter of all deliberations on quantum mechanics.
>
> *(Heisenberg, 1926b, p. 989)*

Whereas Schrödinger hoped that the atomic system would be visualised as a *matter wave* extended in the region of space surrounding the nucleus, the renunciation of the electron orbit in quantum mechanics suggested to Heisenberg that there might exist an analogy between electrons and light quanta. This had been anticipated by Pauli in a letter to Kramers in July 1925, where he wrote: 'in cases where interference phenomena are present, we cannot define definite "trajectories" for the light-quanta' but according to quantum mechanics 'neither can one define any such trajectories for the electrons in an atom'. To this extent Pauli emphasised that 'it can now be regarded as proved that light quanta are just as much (and just as little) physically real as electrons' (Pauli to Kramers, 27 July 1925, pp. 232–5 [item 97]). The problem nevertheless remained: just how was this 'kind of reality' to be conceptualised?

Though critical of the attempts to develop a wave theory of matter, Heisenberg was nonetheless convinced that Schrödinger's wave equation had brought to light the wave–particle duality of the electron, just as Einstein's concept of light quanta had prompted physicists to consider the wave–particle duality of radiation. In a letter to Dirac he stated: 'I see the real progress made by Schrödinger's theory in this: that the same mathematical equation can be interpreted as point-mechanics in a non-classical kinematics *and* as a wave theory according to Schrödinger'. Heisenberg went on to claim: 'I had always hoped that the solution to the paradoxes of quantum theory could later on be found this way' (Heisenberg to Dirac, 26 May 1926, in Mehra & Rechenberg, 2000, p. 202). Again, in his lecture in Düsseldorf in September 1926, Heisenberg explained: 'The extraordinary physical significance of Schrödinger's results lies in the fact that a visualizable interpretation of the quantum-mechanical formulas contains both typical features of a corpuscular theory, as well as typical features of a wave theory' (1926b, p. 993).

It may seem somewhat strange, given Heisenberg's critical reaction to Schrödinger's attempts to develop a wave theory of matter documented in the previous chapter, that in his letter to Dirac in May 1926 he should concede that the same equation of motion could be interpreted *both* as a particle and as a wave theory. Here we should remember that Heisenberg had initially been enthusiastic about Schrödinger's first paper. Writing to Dirac on 9 April 1926, Heisenberg explained that one could 'win a great deal for the physical significance of the theory' from Schrödinger's mathematical treatment of the hydrogen atom (Dirac, 1977, p. 131). But in his letter to Dirac on 26 May, he expressed grave reservations over Schrödinger's efforts to develop a 'wave theory of matter' (Mehra & Rechenberg, 2000, p. 202). In a paper on the many-body problem in quantum mechanics, Heisenberg explained his critique of the wave interpretation of quantum theory:

> Even if a consistent wave theory of matter in the usual three-dimensional space could be developed, corresponding to the programme of Einstein and de Broglie, this would hardly yield an exhaustive description of atomic processes in terms of our familiar space-time concepts. Precisely in view of the emerging close analogy between light and matter one is inclined to believe that such a wave theory of matter is no more a complete description of our atomic experiences than a wave theory of light provides a complete interpretation of optical experiences.
>
> *(Heisenberg, 1926a, p. 412)*

Here, Heisenberg again drew attention to the analogy between light and matter, but did not argue, as Schrödinger had, that what is required in quantum mechanics is a wave theory of the electron. Heisenberg emphasised instead that in quantum mechanics the behaviour of electrons, whether conceived of as waves or as particles, cannot be described in terms of ordinary three-dimensional space, though in his paper Heisenberg does canvass the possibility that we may want to 'consider a space whose structure [*Maßbestimmung*] differs substantially from the Euclidean one, as representing "ordinary" space'

(Heisenberg, 1926a, pp. 412–13). In the absence of any such quantum conception of space and time in the microscopic realm, Heisenberg wrote to Pauli on 23 November 1926 that in quantum theory 'one no longer knows what the words "wave" and "particle" mean' (Pauli, 1979, p. 360 [item 148]).

By 1928, Heisenberg realised that the interpretation of quantum mechanics could be pursued in two separate ways, one taking the 'particle' and the other taking the 'wave' as the fundamental concept. The realisation that these two approaches lead to the same experimental results was, for Heisenberg, critical to the final resolution of the wave–particle dilemma. Our analysis begins with the concept of the particle in quantum mechanics, and the probabilistic interpretation of the wave function.

4.2 Probability waves and the quantum mechanics of particles

As discussed in the previous chapter, one of Heisenberg's criticisms of Schrödinger's attempt at a wave theory of matter in 1926 was that the ψ-function did not, in general, represent a wave-field in ordinary three-dimensional space, but rather a *system* in n-dimensional abstract configuration space where n represents the number of degrees of freedom. Between June and November in 1926, Heisenberg made this point repeatedly in a series of papers, lectures and in his correspondence. In a paper written in June 1926, Heisenberg commented: 'As far as I can see, Schrödinger's method also does not represent a consistent wave theory of matter in the sense of de Broglie' (Heisenberg, 1926a, p. 412). He again made this the focus of his critique of wave mechanics in a lecture on quantum mechanics in September 1926:

> [I]t is possible to establish a wave equation in the coordinate space of 3-f-dimensions for the problem of the motion of f-particles, which then completely replaces the quantum-mechanical problem mathematically ... Until now one has not generally succeeded in directly connecting the Schrödinger wave in phase space with the de Broglie wave in ordinary [three-dimensional] space, which should then be analogous to the light wave. The wave in q-space has therefore until now only a formal meaning.
> *(Heisenberg, 1926b, pp. 992–3)*

Throughout this period, Heisenberg remained convinced that 'one of the most important aspects of quantum mechanics is that it is based on the corpuscular representation of matter', though he was quick to add, 'it does not concern at the same time a description of the motion of the corpuscles in our familiar space-time concepts' (Heisenberg, 1926a, p. 412). The probabilistic interpretation of the ψ-function first proposed by Max Born in his two papers on collisions in June–July

1926 would eventually prove decisive in support of the particle interpretation, though precisely what Born's own views were regarding the reality of waves and particles during this time is far from clear (Born, 1926a, b; Cartwright, 1987). In these papers, Born gave a theoretical treatment of the problem of a collision of an electron with an atom by using the Schrödinger ψ-function. Born was able to represent the collision as a scattering of waves, but he interpreted the square of the wave amplitude of the scattered wave as giving the probability of finding the electron deflected in a particular direction. Here Born introduced for the first time the statistical interpretation of the ψ function. The development of the statistical interpretation of quantum mechanics and its implication for wave–particle duality has been the subject of much historical work though little attention has been specifically devoted to the interpretation developed by Heisenberg (Konno, 1978; Wessels, 1980; Pais, 1982; Beller, 1990).

Though Born's work constituted a definite advance in the treatment of aperiodic processes through wave mechanics, Linda Wessels has rightly pointed out that it was in fact Pauli who first saw the possibility of interpreting the square of the wave amplitude as the probability of finding an electron at a definite position in configuration space in the stationary state of the atom (Wessels, 1980, p. 187). In a letter to Heisenberg on 19 October 1926, Pauli explained that $|\psi(q)|^2 \, dq$ denotes 'the probability that in a definite stationary state of the system the coordinates q of a particle lie between q and $q + dq$' (Pauli, 1979, p. 347 [item 143]). Pauli added: 'From the *corpuscular* standpoint it thus already makes sense for it to lie in multi-dimensional configuration space' (emphasis in original). This interpretation was to prove particularly important for the development of transformation theory by Jordan and Dirac (Dirac, 1927a; Jordan, 1927a). Extending this interpretation to the case of a single free particle in three dimensions, Heisenberg was able to calculate the limits of the accuracy with which we can determine the position and momentum of a particle at any given time. His work marks the first *generalised* probabilistic interpretation of quantum mechanics.

Significantly, Heisenberg's 1927 paper on the interpretation of quantum mechanics takes the concept of the particle as fundamental. Heisenberg explained that if 'we determine the position q of the electron as q' with an uncertainty Δq, then we can express this fact by a probability amplitude which differs appreciably from zero only in a region of spread Δq near q'' (Heisenberg, 1983, p. 69). Writing to Bohr on 19 March 1927, Heisenberg explained that 'one can see that the transition from micro- to macro-mechanics is *now* very easy to understand; classical mechanics is altogether a part of quantum mechanics' (AHQP, emphasis in original). The position and momentum of a point-particle can only be deter-mined within certain limits of precision. In fact, as Heisenberg demonstrated, the more accurately we determine the position of the particle, the less accurately we know the momentum. He was thus able to derive the well-known uncertainty relation for the position and momentum of a particle: $\Delta q\, \Delta p > h/4\pi$.

Although Heisenberg derived the uncertainty relations using Dirac's matrix transformation theory in his 1927 paper, it was clear to him that one could now reinterpret Schrödinger's concept of wave packets. Schrödinger had attempted to describe the classical motion of a particle by assuming that 'material points consist of, or are nothing but, wave-systems' (Schrödinger, 1926, p. 1049). Here he proposed that an electron in a higher energy state of an atom, or a free electron, could be represented as a 'wave packet', or a superposition of waves of different frequencies, which remains localised while moving through space. Yet, as Lorentz pointed out in his correspondence with Schrödinger in 1926, according to classical wave theory 'a wave packet can never stay together and remain confined to a small volume in the long run'. The tracks observed in the Wilson cloud chamber seemed to show that electrons remain localised in the course of their motion, yet as Lorentz explained in his letter to Schrödinger, a wave packet would undergo dispersion 'in the direction of propagation, and even without that dispersion it will always spread more and more in the transverse direction' (Lorentz to Schrödinger, 27 May 1926, Przibram, 1967, p. 47). In response to Lorentz, Schrödinger was able to demonstrate that in the case of the harmonic oscillator a wave packet would stay together (Schrödinger, 1928a). However, this proved to be only an exceptional case. In his paper in 1927, Heisenberg showed that Schrödinger's treatment of the harmonic oscillator was the only case where a wave packet stayed together without dispersing (Heisenberg, 1983, pp. 73–4).

Armed with the probabilistic interpretation of the ψ-function, Heisenberg was able to show that for a high energy state of the atom the wave packet representing the electron's motion inevitably spreads out in space in the course of its periodic motion around the atomic nucleus. But, as he now explained, the wave packet represents the probability of finding the electron at a particular point in space: 'all positions [of the particle] count as likely (with calculable probability) that lie within the bounds of the expanded wave packet' (Heisenberg, 1983, p. 74). The wave packet, therefore, could no longer be thought of as constituting a particle, but must be understood as representing the region of space within which the particle may be located. In other words, the extension of wave packet in space now represented an *uncertainty*, or better, the *indeterminacy*, in the electron's position. In July 1927, Kennard explained that according to Heisenberg's interpretation of the theory, 'the "wave packet" of Schrödinger becomes reinterpreted as the "probability packet"' (Kennard, 1927, p. 326).

This interpretation of quantum mechanics would have radical consequences. Heisenberg now explained, only the 'determination of the position' of the electron through the act of observation 'selects a definite "q" from the totality of possibilities and limits the options for all subsequent measurements'. In this way, 'the results of the later measurements can only be calculated when one again ascribes to the electron a "smaller" wave packet'. Heisenberg now drew the conclusion that the 'probability packet' spreads out in the course of the electron's motion, until the electron interacts with the measuring instrument. The determination of the electron's position 'reduces the wave packet back to its original extension' (Heisenberg, 1983, p. 74). In discussions at the Como conference in September 1927, Heisenberg again emphasised the 'reduction of the wave packet' in describing the motion of a free electron:

> This wave packet moves away not only in a straight line in space, but also it spreads out in the course of time. For a new observation, the wave packet gives the *probability* of finding the electron at a determined position. The new observation itself however reduces the packet to its original magnitude Δq, which makes a selection from a totality of possibilities and thereby reduces the possibilities for the future. This continual change of the wave picture at an observation appears to me a fundamental feature of quantum mechanics. One must just put into practice the conception of *"probability waves"*. The waves do not have immediate reality, of the kind we had earlier ascribed to the waves of Maxwell's theory. One must interpret them as probability waves and therefore expect a sudden change at every observation.
>
> *(Bohr, 1985, p. 140, emphasis added)*

The idea of a 'collapse of the wave packet' in the act of measurement would remain highly controversial, leading to attempts to understand the measurement problem in the 1930s (von Neumann, 1932; London & Bauer, 1983). However, by the late 1920s, this interpretation had become widely accepted, notwithstanding the protestations of physicists such as Einstein and Schrödinger. In September 1927, Ehrenfest (1927) showed that when a particle is acted on by a force, the centre of the wave packet moves according to the classical equation of motion for a point-particle. Soon after, Kennard (1928) and Darwin (1927) undertook the study of the motion of particles using wave mechanics, thus cementing the new interpretation. In their exposition of the new quantum mechanics at the Solvay conference in October 1927, Born and Heisenberg explained how the wave–particle dilemma could be resolved from 'the point of view of the statistical conception of quantum mechanics'. As they now explained: 'The waves are waves of probability' (Born & Heisenberg, 1928, pp. 164–5).

As Heisenberg later explained, while a single electron can be described by a wave (probability) packet in three dimensions, in general the wave equation of quantum mechanics describes a system in abstract multi-dimensional configuration space, not in the ordinary three-dimensional space (Heisenberg, 1958d, p. 43). Friedrich Hund points out, by 1927 Heisenberg, Pauli and Born 'saw the Schrödinger equation as a convenient form of expressing the quantum modification of classical *particle* mechanics' (Hund, 1974, p. 177). Agreement, however, was not universal: a number of physicists were not persuaded that the electron could be described as a 'particle' or a 'material point' since it could not be assigned a well-defined trajectory. In 1931 Schrödinger raised this problem:

> On a little reflection it will be clear that the object referred to quantum mechanics in this connection is not a material point in the old sense of the word. A material point in that sense is a thing situated at a given place ... And if it has a given place at any given moment then surely it must have a definite trajectory, and also, as might be assumed a definite velocity. However, this may be, quantum mechanics forbids the conception of a well-defined trajectory ... We have ceased to believe in the circular and elliptical orbits within the atom. To speak of electrons ... as material points and yet to deny that they have definite orbits appears both contradictory and absurd.
>
> *(Schrödinger, 1957, p. 72)*

Throughout the 1930s, Schrödinger, Planck and von Laue continued to argue that the concept of particle had no physical significance. Quantum mechanics, they argued, could not be regarded as a final or complete theory (Schrödinger, 1934; von Laue, 1932). As Schrödinger later explained, 'to me giving up the path seems giving up the particle' (AHQP, 12 April 1955). Born and Heisenberg were nevertheless adamant that the statistical interpretation of the ψ-function had brought the development of quantum mechanics to a satisfactory conclusion. In their joint paper at the Solvay conference in 1927, they declared: 'we hold the *mechanics of quanta* to be a complete theory, whose fundamental physical and mathematical hypotheses are no longer susceptible to modification' (Born & Heisenberg, 1928, p. 178). Though the electron could not be ascribed a well-defined trajectory, it was still possible, Heisenberg maintained, to speak of the electron as a 'particle', whose motion was described by the laws of quantum mechanics.

In 1926, Heisenberg, it will be remembered, had foreshadowed the possibility of *both* a (non-classical) particle mechanics and a (non-classical) wave theory. There is more, then, to Heisenberg's conception of the physical reality of the 'waves' than the probabilistic interpretation of the wave function. A deeper understanding of Heisenberg's view of the wave–particle dilemma is only possible once we examine the early development of quantum electrodynamics in 1927–8. As Heisenberg commented in 1933, whereas 'Schrödinger's theory is concerned with waves in multidimensional space', the papers written by Jordan, Klein and Wigner on quantum electrodynamics showed that it was indeed possible to develop a wave theory of matter in three-dimensional space (Heisenberg, 1985a, p. 230). It is worth noting here that, notwithstanding the work of Hund (1974, pp. 177–81), Bromberg (1977), Darrigol (1986), and more recently Mehra and Rechenberg (2001, pp. 191–232), this phase of quantum electrodynamics has received relatively little attention from historians of science. Yet, as we shall see, for Heisenberg, this development was of critical importance in resolving the problem of the wave–particle duality. It is to this that we now turn our attention.

4.3 Quantised matter waves and wave–particle equivalence

In February 1927, Bohr delivered a paper in Copenhagen, written by Dirac, entitled 'The Quantum Theory of the Emission and Absorption of Radiation'. It was to have a huge impact on the development of quantum electrodynamics. In it Dirac proposed a method for *quantising* the electromagnetic wave, which was later termed 'second quantisation'. By treating the light wave as a 'quantised wave', Dirac was able to show that one could obtain the same results as could be obtained by using 'the light-quantum treatment' (Dirac, 1927b, p. 265). As Oliver Darrigol explains, in Dirac's paper on wave quantisation, 'one starts from the classical wave picture of electromagnetic radiation, applies to it the formal rules of quantization, and gets the discontinuous structure of radiation energy' (Darrigol, 1986, p. 228). Dirac himself pointed out that through wave quantisation it was now possible to understand how 'the wave point of view is consistent with the light quantum point of view' (Dirac, 1927b, p. 265). A major breakthrough had thus occurred in the understanding of the wave–particle duality of light.

Dirac's method was to provide the inspiration for a new advance in the development of a wave theory of *matter*. In the summer of 1927, Jordan appropriated the method of second quantisation and applied it to the wave equation for multiple electrons in three-dimensional space. In July he wrote to Schrödinger, explaining that it had occurred to him that by quantising the electron's wave-field, 'a complete theory of light and matter could be derived in which, as an essential ingredient, this wave field operates in a quantum non-classical way' (Schweber, 1994, p. 34). Jordan made this idea public in a paper on gas degeneracy in July 1927, in which he explained that it was now possible to represent electrons, analogously to light quanta, 'by *quantised waves in conventional three-dimensional space*' (Jordan, 1927c, p. 473, emphasis in original). Jordan concluded:

a quantum-mechanical wave theory of matter can be developed that represents electrons by quantised waves in the usual three-dimensional space. The natural formulation of the quantum theory of electrons will be attained by conceiving of light and matter as interacting waves in three-dimensional space. The basic fact of electron theory, the existence of discrete electric particles, appears in this context as a characteristic of quantum phenomena, namely to be equivalent to the fact that matter waves appear only in discrete quantum states.

(Jordan, 1927b, p. 480)

In October 1927, Jordan and Klein (1927) published a significant paper in which they were able to derive the wave equation for more than one particle by using Dirac's method of quantisation starting from a classical wave equation for electrons in three-dimensional space. Jordan and Wigner extended this interpretation in an important paper in which they developed further the idea of quantised matter waves. This work made it possible to formulate the interaction of the electron wave-field with the electromagnetic field in a way 'which avoids the wave representation in the abstract [multidimensional] coordinate space in favour of a representation by quantum waves in the usual three-dimensional space'. An explanation was now available for 'the existence of material particles ... similar to [Dirac's explanation of] the existence of light quanta', namely through the quantisation of waves (Jordan & Wigner, 1928, p. 631).

This crucial breakthrough in the quantum theory of the electron paved the way for a translation of Schrödinger's wave equation in configuration space into a wave theory of matter in three-dimensional space. However, these 'waves' were not to be understood in the classical sense. Just as Heisenberg had quantised the classical equation of motion for a particle in 1925 (with the result that the particle no longer possessed a well-defined position or momentum), so Jordan quantised the classical wave equation (with the result that the wave amplitude no longer assumed a well-defined value). Heisenberg drew the following conclusion in 1929: 'it follows that there also exist *indeterminacy relations* for the wave picture, that can be derived mathematically from the commutation relations of the wave amplitudes' (Heisenberg, 1929, p. 494, emphasis in original). As Born would put it later, when the three-dimensional waves are quantised, 'the statistical character of the ψ-function is introduced at a deeper and more abstract level' (Born, 1956a, p. 143).

Although Heisenberg had throughout 1926 rejected the possibility of a *classical* wave interpretation of quantum mechanics along the lines set out by de Broglie and Schrödinger, he responded to the Klein–Jordan theory of *quantised* matter waves with enthusiasm. We have already seen that one of Heisenberg's key objections to Schrödinger's original wave interpretation was that the wave equation only gave a representation in configuration space of many dimensions.

As Heisenberg had repeatedly stressed, it was not yet possible to give a de Broglie wave interpretation in ordinary three-dimensional space. Yet, Heisenberg had foreshadowed such a possibility in his earlier discussion of quantum electrodynamics. His anticipation of Jordan's work on the quantised matter wave is perhaps most clearly expressed in a letter to Pauli in February 1927:

> I agree very much with your program concerning electrodynamics, but not quite concerning the analogy: quantum-wave-mechanics: classical [particle] mechanics = quantum electrodynamics: classical Maxwell [wave] theory. That one must quantise the Maxwell equations to get light quanta and so on *à la* Dirac, I believe already; but perhaps the de Broglie waves will later also be quantised in order to obtain charge and mass and statistics (!!) of electrons and nuclei.
>
> *(Heisenberg to Pauli, 23 February 1927, Pauli, 1979, p. 376 [item 154])*

We may reasonably infer from this passage that Heisenberg did not share Pauli's view that quantum mechanics is inextricably connected to the concept of a point-mass, while quantum electrodynamics is based on the concept of the electromagnetic field. Heisenberg entertained the possibility that quantum mechanics could be derived from the quantisation of matter waves. This possibility emerges again in the report he and Born presented at the Solvay conference in October 1927 (Born & Heisenberg, 1928, p. 144), but was finally realised only with the publication of Jordan–Klein and Jordan–Wigner papers later in 1927–8. As Heisenberg explained to Kuhn in an interview in 1963, 'I was somewhat excited by the idea of quantizing the Schrödinger waves but that was on account of the problem of interpretation' (AHQP, 12 July 1963, p. 4). For this reason, Heisenberg claimed he had 'found these several papers of Klein–Jordan–Wigner extremely satisfactory' (AHQP, 12 July 1963, p. 5). The Jordan–Klein paper, in particular, was deemed 'an extraordinarily important work', not only because it paved the way for quantum electrodynamics, but also because it shed new light on the wave–particle duality of matter (AHQP, Heisenberg to Bohr, 5 December 1927).

There was, for Heisenberg, an important difference between the Schrödinger waves and quantised waves of Jordan, Klein and Wigner. The Schrödinger wave equation could only be interpreted as a probability wave in abstract configuration space, and therefore presumed the existence of particles. On the other hand, the wave theory of matter developed by Jordan, Klein and Wigner was not to be interpreted in this way. Here, the wave amplitudes were not to be interpreted as probability amplitudes, but rather as representing 'real' matter-wave fields in three-dimensional space. Thus from the perspective of the quantum electrodynamics of Jordan, Klein and Wigner, the concept of the wave was, ontologically speaking, more fundamental than the concept of the particle. In his 1955–6 Gifford lectures, Heisenberg explained:

As early as 1928 it was shown by Jordan, Klein, and Wigner that the mathematical scheme can be interpreted not only as the quantisation of particle motion but also as a quantisation of three-dimensional matter waves; therefore, there is no reason to consider theses matter waves as less real than the particles ... only the waves in configuration space (or the transformation matrices) are probability waves in the usual interpretation, while the three-dimensional matter waves or radiation waves are not. *The latter have just as much and just as little "reality" as the particles*; they have no direct connection with probability waves but have a continuous density of energy and momentum, like an electromagnetic field in Maxwell's theory.

(Heisenberg, 1958d, pp. 118, 126, emphasis added)

This passage makes it clear that for Heisenberg the three-dimensional matter waves, unlike the probability waves in the usual interpretation, were to be understood as 'real' waves, in the same way that the 'particle' in the probabilistic interpretation was understood as 'real'. This marked a shift from the position he had taken at the Como conference, where he argued that the 'waves do not have immediate reality, of the kind we had earlier ascribed to the waves of Maxwell's theory'. Writing to Bohr, Heisenberg declared, 'I now believe that the fundamental questions are completely solved' (AHQP, 23 July 1928), a conclusion which he reaffirmed in his interview with Kuhn: 'I made some effort to explain to Bohr that these papers of Klein–Jordan–Wigner were just a very good illustration of what he wanted with the complementarity because there was a complete symmetry between waves and particles'. However, as Heisenberg recalls, Bohr at first 'did not feel exactly that way' though precisely what Bohr's views at this time were is difficult to judge (AHQP, 12 July 1963, p. 6). As we shall see below, Bohr's understanding of wave–particle duality differed substantially from Heisenberg's.

With the emergence of the theory of quantised matter waves, it was now possible to see that one could arrive at precisely the same equation of motion by a quantum modification of *particle* mechanics through the statistical interpretation of the Schrödinger equation or by a quantum modification of three-dimensional *wave* theory through quantum electrodynamics. This is the key to Heisenberg's claim in 1929 regarding 'the *complete equivalence* of wave and particle pictures in quantum mechanics' (Heisenberg, 1929, p. 494). In the appendix of his Chicago lectures in 1929, Heisenberg argued that in quantum theory 'the particle picture and the wave picture are merely two different aspects of one and the same physical reality' in the sense that 'one and the same set of mathematical equations can be interpreted at will in terms of either picture' (Heisenberg, 1930, pp. 177–8). This is precisely what Heisenberg had anticipated in his letter to Dirac in May 1926. In his interview with Kuhn in 1963, he explained how he had presented the problem:

I wrote these lectures which I gave in 1929. There I think it was always done with the two possibilities. It was not the dualism in the sense that you needed both, it was rather a dualism in the sense that you may do it either way. You can either start from the wave picture [Jordan–Klein–Wigner] or you can start from the particle picture [Born]. Each of the two pictures can be carried through to the end [or quantised] and will give you correct answers. At the end you discover that, after all, you have done the same thing, just in a different language. I think that these lectures at least were clearly written in the sense that they showed not that we have a dualism, but that you have the two possibilities.

(AHQP, 23 February 1963, p. 20)

As the passage quoted above suggests, Heisenberg did not think that in quantum mechanics the physicist is confronted with a wave–particle duality in the sense that there exist both matter waves and material particles, as for example de Broglie argued in 1927, but rather the *equivalence* of two modes of description. As he later put it, whether we take the electron to be a particle or a wave, in quantum mechanics, 'we have to quantize both and then they are the same thing' (AHQP, 12 July 1963, p. 6). In describing the stationary state, for instance, we can employ either the language of *particles*, according to which the ψ-function describes the probability of finding an electron at a point in space, or alternatively, the language of *wave theory*, according to which we speak of the distribution of charge density surrounding the nucleus. The 'electron' can therefore be treated as a particle, whose position and momentum can be determined only within certain limits of accuracy, or conversely, as a spatially extended, quantised wave-field in three-dimensional space.

4.4 Heisenberg and Bohr: divergent views of wave–particle duality

Heisenberg's conception of the wave–particle duality was closely linked with the mathematical physics of Born, Pauli, Jordan and Klein. Bohr's views, on the other hand, were shaped more by his preoccupation with experimental phenomena such as the interference and diffraction of electrons demonstrated by Davisson and Germer in April 1927, which appeared to reveal the wave-like nature of matter. In other words, Bohr was concerned primarily with the *experimental phenomena* in his notion of wave–particle duality, whereas for Heisenberg, the duality was illustrated through the different transformations of the *fundamental equation* of quantum mechanics. As already mentioned the wave–particle duality was the subject of intense discussions between Bohr and Heisenberg in the spring of 1927, during which time Bohr formulated his notion of 'complementarity'. The debate between Bohr and Heisenberg in the spring of 1927 is often thought to have finally convinced Heisenberg of the importance of Bohr's notion of wave–particle complementarity in quantum theory (Jammer, 1974, p. 69). However, this is not so. As we shall see, the wave–particle duality continued to be the source of much disagreement and misunderstanding well beyond their discussions in 1927.

As Heisenberg later recalled, it emerged in the course of these discussions that 'Bohr sought to take the duality between the wave picture and the corpuscular picture as the starting point of the physical interpretation' (AHQP, 28 February 1963, p. 10). In his later accounts, Heisenberg stressed that by the spring of 1927 'there remained conceptual differences' between himself and Bohr, though these 'just referred to the different starting points or to the different ways of expressing things, but no longer to a different interpretation of the theory' (Heisenberg, 1960a, pp. 46–7). In 1928, Oscar Klein, who was present during the discussions in Copenhagen, explained that the dispute between Bohr and Heisenberg had revolved around whether wave–particle complementarity or discontinuity was more fundamental in quantum theory (AHQP, BSC, Klein to Birtwhistle, 3 May 1928). Heisenberg's letter to Pauli on 16 May 1927 confirms this account. There, Heisenberg expressed a somewhat critical attitude towards Bohr's new viewpoint:

> [W]e have here discussed quantum theory at length. Bohr plans to write a general treatise on the "conceptual structure" of quantum theory, from the viewpoint that "there exist waves and particles" – if one begins at once with that one can naturally make everything contradiction free. In respect to this work, Bohr has drawn my attention to the fact that ... certain points could be better expressed and discussed in every detail, if only one begins with a quantitative discussion directly with the waves. Nevertheless, I am naturally now as before of the opinion that the discontinuities are the only interesting things in quantum theory and that one can never stress them enough.
>
> *(Pauli, 1979, pp. 394–5 [item 163])*

While a compromise of sorts was reached by Pauli in reconciling the views of Bohr and Heisenberg in May 1927, the 'note in proof' added to Heisenberg's paper on the uncertainty relations did not, as many scholars have argued, signify that the two physicists had reached complete agreement (Stepansky, 1997, p. 384). As Heisenberg would later recall: 'The main point was that Bohr wanted to make this dualism between waves and corpuscles as the central point of this problem and to say, "That is the center of the whole story, and we have to start from that side of the story in order to understand it"' (AHQP, 25 February 1963, p. 18). Heisenberg had a different view: 'for me it was clear that ultimately there was no dualism' (in the sense that it was necessary to employ both wave and particle concepts in the description of one and the same object), and 'therefore I was always a bit upset by this tendency of Bohr of putting it into a dualistic form' (AHQP, 5 July 1963, p. 11). While Heisenberg's later writings may give the impression he had reconciled his differences with Bohr, a closer reading of their respective texts tells a different story.

In the paper on complementarity published in 1928, Bohr was adamant that in confronting the wave–particle duality in quantum theory 'we are not dealing with contradictory but with complementary pictures of the phenomena, which *only together* offer a natural generalization of the classical mode of description' (Bohr, 1928, p. 581, emphasis added). To this extent, Bohr argued that it was *necessary* to use both wave and particle concepts in quantum theory. Heisenberg, however, seems to have adopted a different view. In 1929, he pointed out that while it was certainly possible to determine the limits of the concept of a particle in quantum theory by recourse to the concept of wave packets, he maintained it was also possible to do so 'without explicit use of the wave picture'. The uncertainty relations for the electron's position and momentum, Heisenberg argued, 'are readily obtained from the mathematical scheme of quantum theory and its physical interpretation' (Heisenberg, 1930, pp. 15–16). There is no need to invoke the wave picture of matter.

A closer examination of Heisenberg's later recollections of his discussions with Bohr on this point lends further support to the view that Heisenberg never accepted Bohr's version of the wave–particle duality. In his interview with Kuhn, Heisenberg conceded that in his conversations with Bohr he had agreed 'it was *convenient* also to speak about waves' in describing the motion of electrons, though he maintained, 'it was *not essential* to do it' (AHQP, 25 February 1963, p. 20, emphasis added). One could just as easily restrict oneself to speaking about the motion of a *particle* in quantum mechanics. Indeed, Heisenberg explained that in actual fact he had never fully accepted the view that it was absolutely necessary to use both wave and particle concepts, though he conceded that it was perhaps useful for physicists to think in these terms. He recalled the critical period in 1927 during which time he came to a kind of compromise with Bohr:

> I could perhaps say that at that time I had understood that it doesn't do any harm to my own explanation if I do it that way. I felt, "Well, one might also do it that way. Why not?" Therefore I didn't want to protest too strongly against it. I said, "All right, it may be of some help to play always between both pictures". For me the essential point was that I had understood that by playing between the two pictures, nothing could go wrong. So I didn't object to playing with both pictures. At the same time I felt that it was not necessary. I would say it was possible but not necessary.
> *(AHQP, 25 February 1963, p. 21)*

While Heisenberg often seems to have endorsed Bohr's view of wave–particle duality in the 1930s, substantial differences remained that were never fully resolved. In Heisenberg's hands, the wave–particle complementarity was transformed from a general philosophical viewpoint into a pedagogical principle. He told Kuhn in 1963: 'For explaining the gamma ray microscope to physicists, it was useful to play between both pictures. That I could see. But it was not absolutely essential' (AHQP, 25 February 1963, p. 21). In Heisenberg's mind it was possible to speak exclusively in terms of light quanta or alternatively in terms of electromagnetic waves. In this way, beneath the veneer of agreement, Bohr and Heisenberg held quite different views of the dual nature of light and matter.

Although somewhat equivocal about the need to employ both wave and particle descriptions of the behaviour of electrons in quantum theory, in his 1929 Chicago lectures, Heisenberg did draw attention to the fact that the 'wave' or 'particle' features of matter and radiation were brought out sharply by different experimental arrangements. On the one hand, the Wilson photographs of electrons in a cloud chamber, the scattering of electrons in the Compton–Simon experiment and the collision experiments carried out by Franck and Hertz suggested that electrons were particles. On the other hand, the diffraction of electrons carried out by Davisson, Germer, Thomson and Rupp in 1927–8 seemed to provide good evidence for the existence of matter waves. In perhaps the most elaborate statement of the wave–particle duality to appear from the late 1920s, Heisenberg concluded his introductory chapter to *The Physical Principles of the Quantum Theory* in typical Bohr-like fashion:

> From these experiments it is seen that both matter and radiation possess a
> remarkable duality of character, as they sometimes exhibit the properties of waves,
> at other times those of particles. Now it is obvious that a thing cannot be a form of
> wave motion and composed of particles at the same time – the too concepts are too
> different ... As a matter of fact, it is experimentally certain only that light sometimes
> behaves as if it possessed some of the attributes of a particle, but there is no
> experiment which shows that it possesses all the properties of a particle; similar
> statements hold for matter and wave motion. The solution of this difficulty is that the
> two mental pictures which experiments lead us to form – the one of particles, the other
> of waves – are both incomplete and have only the value of analogies which are
> accurate only in limiting cases ... yet they may be justifiably used to describe things
> for which our language has no words. Light and matter are both single entities, and
> the apparent duality arises in the limitations of our language.
>
> *(Heisenberg, 1930, p. 10)*

While this passage suggests that by 1929 Heisenberg had largely accepted Bohr's view of wave–particle duality, there were important differences. While it was apparent that an electron 'sometimes behaves as if it possessed some of the attributes of a [classical] particle', Heisenberg was careful to point out that 'there is no experiment which shows that it possesses all the properties of a particle' (i.e. it possesses a simultaneously well-defined position and momentum). Moreover, one could, and sometimes did, employ both wave and particle descriptions in the *same* experiment. Describing Bohr's analysis of the gamma-ray microscope thought-experiment, Heisenberg again emphasised '*simultaneous* use is made of deductions from the corpuscular and wave theories of light' (Heisenberg, 1930, pp. 22–3). Such a view, however, is at odds with the view Bohr expressed in his letter to Einstein on 13 April 1927, in which he argued that the concept of light quanta could be brought 'into harmony with the consequences of the wave theory of light' only because 'the different aspects of problem *never appear at the same time*' (Bohr, 1985, p. 22, emphasis added). Even when Heisenberg defended Bohr's view that wave and particle pictures were 'complementary' to the extent that they were to be employed in mutually exclusive experimental conditions, he remained equivocal, restricting himself to the more moderate view that for certain experiments '*it might be more convenient* to imagine that the atomic nucleus is surrounded by a system of stationary waves' rather than to visualise the electron as a particle moving in a well-defined trajectory around the nucleus (Heisenberg, 1958a, p. 40, emphasis added). The equivalence of the two transformations of the quantum-mechanical formalism showed that the physical interpretation of experiments could be carried out wholly in terms of a (quantised) wave theory or alternatively in terms of a (quantised) particle theory.

Though much has been made of Bohr's notion of wave–particle duality, Bohr actually devoted very little attention to it in his published writings, and what little he did write on the subject is ambiguous. Indeed, while it is often contended that the wave–particle duality is one form of complementarity, Bohr actually rarely, if ever, presented it as such. As Max Born pointed out, 'One could still call the use of particles and waves in physics a duality in the description' but this 'should be strictly distinguished from complementarity' (Born, 1956b, p. 106). It is widely accepted that after 1927, Bohr's interpretation rested on the view that both 'wave' and 'particle' pictures were necessary for a full account of experimental phenomena. However, on some occasions Bohr appears to have argued it was not possible to directly observe 'light quanta' or 'matter waves', and to this extent both must be regarded merely as 'symbols helpful in the formulation of probability laws' (Bohr, 1932, p. 370). In a seldom-quoted lecture in 1930, Bohr expressed this view:

> In this sense, phrases such as the "corpuscular nature of light" or the "wave nature of electrons" are ambiguous, since such concepts as corpuscle and wave are only well defined within the scope of classical physics, where, of course, light and electrons are electromagnetic waves and material corpuscles respectively.
>
> *(Bohr, 1932, p. 370)*

Bohr presented a similar view in his lecture on 'Maxwell and modern theoretical physics' delivered in 1931, in which he argued that *even in quantum mechanics* we must not forget that 'only the classical ideas of material particles and electromagnetic waves have a field of unambiguous application, whereas the concepts of photons and electron waves have not' (Bohr, 1931, p. 691). Dugald Murdoch, whose study of Bohr's philosophy represents by far the most thorough investigation of Bohr's views on this question, argues that these passages, while seldom referred to, provide us with a deeper insight into Bohr's view of wave–particle duality. Murdoch concludes that the idea that the electron is fundamentally a 'particle' and that light is fundamentally a 'wave' represents Bohr's most carefully considered position on the wave–particle duality – a position which finds its clearest expression in Bohr's writings after 1929 (Murdoch, 1987, pp. 67–79). If we accept Murdoch's reading of Bohr, which draws on both his published and unpublished writings, we can conclude that Bohr and Heisenberg held sharply diverging views on the 'reality' of waves and particles.

Murdoch's reading of Bohr is supported by the accounts given by two physicists with whom Bohr worked closely – Oscar Klein and Léon Rosenfeld. In an interview with Kuhn, Klein recalled that in discussions on complementarity in 1927, Bohr emphasised that it was necessary to use the concept of matter waves in describing electrons, but he always did so 'with the knowledge that these were not waves in the literal sense, and he pointed out very strongly also that in the literal sense one might say that electrons are particles and that the electromagnetic waves are waves' (AHQP, 16 July 1963, p. 7). Rosenfeld, who worked closely alongside Bohr during the 1930s, has also argued that in considering the wave–particle duality of light, Bohr was inclined to see 'the electromagnetic field as being in some sense more fundamental than the photon concept'. While there is general agreement that Bohr held such a view prior to 1927, Rosenfeld contends that: 'This is a point of view that Bohr never abandoned' (Rosenfeld, 1979b, p. 690). According to Rosenfeld, Bohr never accepted the symmetry of wave and particle descriptions of matter either – in the case of the electron he held the particle concept to be more fundamental, 'whereas the wave aspect is the symbolic one' (Rosenfeld, 1979b, p. 700). The accounts provided by Klein and Rosenfeld are significant, not only because they give us reason to question the standard view of complementarity, but also because they attribute to Bohr a view which stands in sharp contrast to Heisenberg's notion of wave–particle equivalence.

While there remains some conjecture as to precisely what Bohr's views were concerning the 'reality' of particles and waves, it is clear that he never attributed the same importance to the Jordan–Klein–Wigner papers on wave quantisation as Heisenberg did. In 1955, Heisenberg credited Jordan, Klein and Wigner with having 'demonstrated for the first time the complete *equivalence* of wave and particle pictures in the quantum theory' (Heisenberg, 1955, p. 15). Indeed, by the 1950s Heisenberg would reinterpret Bohr's original notion of wave–particle complementarity through his own quite distinct notion of wave–particle equivalence. In his Gifford lectures in 1955–6, Heisenberg again argued, 'the dualism between two complementary pictures – waves and particles – is also clearly brought out' by the possibility of interpreting the quantum-mechanical formalism in two different ways. While the Schrödinger wave equation is normally written to express the probability of finding 'the co-ordinates and the momenta of the particles' in n-dimensional configuration space, 'by a simple transformation it can be rewritten to resemble a wave equation for an ordinary three-dimensional matter wave'. The fact that we can regard the electron as *either* a wave or a particle, Heisenberg insisted, 'does not lead to any difficulties in the Copenhagen inter-pretation' (Heisenberg, 1958d, pp. 50–1).

Rather than use *both* classical pictures in a complementary description, or in mutually exclusive experimental arrangements, Heisenberg argued that one could interpret the quantum equation of motion for an electron *either* as a quantised wave in three-dimensional space *or* as describing the quantised motion of a particle. For Heisenberg, the emergence of the new quantum electrodynamics gave formal expression to what he termed the 'complete equivalence of wave and particle pictures in quantum theory'. The distinction to be drawn between Heisenberg's equivalence and Bohr's wave–particle duality was disguised by the fact that in later years Heisenberg frequently used the term 'complementarity' to express his own notion of equivalence.

Although the Copenhagen interpretation is virtually synonymous with Bohr's notion of complementarity, Heisenberg pursued his own interpretation between 1926 and 1928, which reflected both the probabilistic interpretation of the ψ-wave, to which he himself had contributed decisively through the notion of the probability packet, and the 'quantisation' of the three-dimensional wave carried out by Dirac, Jordan, Klein and Wigner in 1927–8. According to Heisenberg, these two interpretive frameworks constituted the two sides of the wave–particle dilemma. In the first case, the wave-like phenomena such as diffraction and interference could be explained by treating electrons as particles. In the second case, the corpuscular nature of electrons is a manifestation of the quantisation of matter waves. As a consequence, quantum mechanics could be interpreted either way, through two quite different ontologies.

Heisenberg's view of wave–particle duality is significant for two reasons. First, because it offers an original and unique insight into the problem of wave–particle duality, which to this day stands as one of the central paradoxes of quantum mechanics. And secondly, as it turns out, Heisenberg's notion of wave–particle equivalence differed quite markedly from Bohr's notion of wave–particle complementarity. This divergence between the two men calls into question the standard historical view that after a brief period of intense disagreement in spring 1927, Heisenberg accepted Bohr's complementarity argument.

It is true that Bohr and Heisenberg both accepted some form of wave–particle duality. However Bohr's emphasis on the complementary use of wave and particle descriptions arose from his preoccupation with the *experimental object* of quantum mechanics. The electron exhibits both wave-like phenomena of interference and diffraction, and particle-like phenomena in different experimental situations. Heisenberg, on the other hand, saw the possibility of interpreting the object of quantum mechanics as a wave or as a particle, because of the different transformations of the quantum-mechanical equation of motion. To this extent, his understanding of the duality arose from the possibility of interpreting the *mathematical formalism* in two different, but empirically equivalent, ways. In highlighting the hidden disagreement between Bohr and Heisenberg, this chapter serves to disentangle the different viewpoints which remained in tension within the Copenhagen interpretation.

While physicists like de Broglie and Einstein saw the resolution of the wave–particle duality as being of fundamental importance in the interpretation of quantum mechanics, Heisenberg saw it as a manifestation of a deeper problem – namely the inadequacy of classical words like 'wave' and particle'. In the passage quoted earlier, he expressed the view that 'the apparent duality arises in the limitations of our language' (Heisenberg, 1930, p. 10). In an interview with Kuhn, Heisenberg again explained that 'the fact that we can use two kinds of words [like wave and particle] to describe quantum mechanics is just an indication of the inadequacy of words' (AHQP, 25 February 1963, p. 20). Indeed, as we shall see, the limits of classical concepts, and the language-reality problem, would become the central theme in Heisenberg's philosophy of quantum mechanics. Heisenberg's philosophical concerns with language first became apparent in his conception of quantum indeterminacy, which forms the subject of the next chapter.

5

Indeterminacy and the limits of classical concepts: the turning point in Heisenberg's thought

The publication of Heisenberg's paper *Über den anschaulichen Inhalt der quantentheoretische Kinematik und Mechanik* in March 1927 marks one of the turning points in the interpretation of quantum mechanics (Heisenberg, 1927a). In the paper, Heisenberg argued that the accuracy with which we can know both the position and momentum of a particle, such as an electron, is subject to an in principle limitation. The more precisely we can determine the particle's position, the less precisely we can know its momentum, and vice versa. Heisenberg was able to derive the well-known mathematical relation for the product of the uncertainties of the position and momentum of a free particle $\Delta p.\Delta x \geq h/2\pi$ using the transformation theory developed by Jordan and Dirac.[1] The epistemological meaning of the uncertainty or indeterminacy principle, as it is commonly known, was the subject of much debate among physicists and philosophers in the 1930s (Schrödinger, 1930; von Laue, 1934; Cassirer, 1956; Schlick, 1979c). Indeed, a number of different philosophical interpretations of the uncertainty principle emerged in the 1930s. Some interpreted it as a statistical scatter relation, others saw it as an expression of the limits of our capacity to measure the exact position and momentum of a particle, and still others saw it as an essential feature of the particle itself (McMullin, 1954; Jammer, 1974, pp. 56–84; Uffink & Hilgevoord, 1985). The meaning of the uncertainty relation would become central to the debates over the interpretation of quantum mechanics.

But what exactly was Heisenberg's philosophical view of the indeterminacy principle in quantum mechanics? Historical scholarship into Heisenberg's formulation of the uncertainty principle has recently emerged from the work of Mara Beller (1985, 1999, pp. 65–101) and Melanie Frappier (2004). Beller, in

---

[1] Hilgevoord and Uffink have also conducted a thorough examination of the mathematical foundations of the uncertainty relations more generally, not only of position and momentum, but also other non-commuting quantities in quantum mechanics like energy and time (Hilgevoord & Uffink, 1988, 1990).

particular, has emphasised that at the time Heisenberg was formulating his thoughts on indeterminacy, his own philosophical view was in a state of transition. In this chapter, I take up this notion by focusing exclusively on one of the central themes of the uncertainty paper – the limited applicability of classical concepts in quantum theory. My aim is to show that Heisenberg's conceptualisation of 'indeterminacy' passes through three different, but over-lapping, phases in the critical period between 1926 and 1928, each of which constituted a distinctive approach to the development of a new conceptual framework for quantum mechanics.

After the development of matrix mechanics in 1925, Heisenberg became convinced that quantum mechanics demanded a fundamentally new conception of space and time. This was made explicit in his correspondence with Pauli in the second half of 1926, during which time Heisenberg expressed the view that space and time may be statistical concepts, or alternatively that space-time might have a discontinuous structure. Nevertheless, by February 1927 he had abandoned this line of thought and instead turned his attention to a redefinition of the concepts of position and velocity through an operational analysis.[2] This approach, which is the one we find in the uncertainty principle paper, represents a conscious attempt on Heisenberg's part to employ the same method of analysis to quantum mechanics as Einstein had in redefining the concept of simultaneity in the special theory of relativity. However, Heisenberg's commitment to this operational viewpoint was short-lived. After fractious discussions with Bohr in 1927, Heisenberg gradually came to realise that this method of defining con-cepts was untenable, and he subsequently abandoned the operational theory of meaning which had characterised his 1927 paper. This realisation signifies an important turning point in Heisenberg's philosophy, in which he abandoned a verification theory of meaning, and finally accepted Bohr's doctrine of the indispensability of classical concepts.

In bringing to light the shift that occurs in Heisenberg's thinking during this period, I seek to draw attention to two important points which have often been ignored or misunderstood in the secondary literature. First, Heisenberg's intro-duction of the imaginary gamma-ray microscope was not intended primarily to demonstrate the limits of precision in measurement. Though it certainly did this, its real purpose was to define the concept of position through an operational analysis. This becomes evident once we situate Heisenberg's use of an imagi-nary gamma-ray microscope within the context of his concerns over the

---

[2] It should be noted that Heisenberg did return to the idea of discontinuous space-time in 1930, but these later speculations must be seen in the context of his efforts to build a new quantum field theory, and did not form part of his understanding of quantum mechanics, with which we are concerned here (Carazza and Kragh, 1995).

meaning of concepts in quantum theory. Much of the discussion of this issue has traditionally focused on whether Heisenberg thought the uncertainty relations represented a limit of our *knowledge* of the precise position and momentum of a particle, or whether he was inclined to interpret this relation as indicating that the *electron itself* did not possess a well-defined position and momentum (Roychoudhuri, 1978; Cassidy, 1998). It seems clear, however, that in spite of the ambiguous way in which Heisenberg sometimes expressed himself, he never subscribed to the view the uncertainty relations merely represent our ignorance of the electron's precise position and momentum at a given time. We can make better sense of why Heisenberg chose to express himself as he did once we realise that his concerns at the time of writing his paper were not primarily *epistemic*, but rather *semantic*.

Secondly, I offer a different perspective on the Bohr–Heisenberg dialogue in Copenhagen in 1927 to the one usually presented. Much of the literature takes the view that the discussions between Bohr and Heisenberg over the gamma-ray microscope centred on the significance of wave–particle duality. This is primarily because the undulatory and corpuscular features of light and matter appear to have been Bohr's central concern at this time, in contrast to Heisenberg's emphasis on discontinuity (Jammer, 1974, p. 66; Stepansky, 1997; Tanona, 2004). As I argued in the previous chapter, however, Bohr and Heisenberg never reached agreement on this issue. Yet, the discussions with Bohr did leave a lasting impression on Heisenberg for another reason – they convinced him that classical concepts such as 'position' and 'momentum' were indispensable in quantum mechanics despite the limitations on their applicability. This marks a crucial turning point in Heisenberg's thought. In order to see more clearly the overall shift in Heisenberg's thinking, it is necessary to turn first to his attempts to find an interpretation of quantum concepts through new concepts of space and time.

5.1 New concepts of space and time in quantum theory

As we have already seen, in the period between 1925 and 1927, during which time quantum mechanics was developed, Heisenberg had abandoned the view that the electron was a point-particle in the classical sense. Heisenberg, along with Born and Jordan, argued that 'the motion of electrons cannot be described in terms of the familiar concepts of space and time' (Born, Heisenberg & Jordan, 1967, p. 322). Though there was disagreement on whether the electron was in reality a particle or a wave, a number of leading physicists in Copenhagen and Göttingen took the view that quantum mechanics would require an entirely new conceptual framework, in which the basic concepts of classical particle physics, such as 'position' and 'velocity', would no longer occupy a central place. This view continued to be expressed by physicists such as Max von Laue and Erwin Schrödinger well into the 1930s (Beller, 1988, pp. 159–61). Writing in 1936, the philosopher Ernst Cassirer gave clear expression to this line of thought:

> We must squarely face the problems thus created. There seems to be no return to the lost paradise of classical concepts; physics has to undertake the construction of a new methodological path ... If it appears that certain concepts, such as those of position, velocity, or of the mass of an individual electron can no longer be filled with definite empirical content, we have to exclude them from the theoretical system of physics, important and fruitful though their function may have been.
>
> *(Cassirer, 1956, pp. 194–5)*

The view here expressed by Cassirer stands in sharp contrast to the interpretation of quantum mechanics developed by Bohr, Heisenberg and Pauli in the late 1920s, according to which classical concepts such as position and momentum remain indispensable – even in quantum mechanics, where they have only limited applicability. Yet, if we look closely at Heisenberg's efforts to develop new conceptual foundations for quantum mechanics *prior to 1927*, we find that his point of view has much in common with the one expressed in the passage quoted above. Indeed, the attempt to find new quantum concepts was integral to Heisenberg's project of reconstructing the foundations of quantum mechanics between 1925 and 1927. As his correspondence with Pauli during this period shows, Heisenberg's early efforts to resolve the difficulties associated with quantum theory took the form of an attempt to construct a new conception of space and time at the sub-atomic level.

As we have already seen, Heisenberg had repeatedly emphasised the restrictions imposed by quantum theory for a space-time description. In his paper on the many-body problem published in June 1926, he argued that while electrons behave as thought they were particles in the sense that they possess a definite mass and electric charge, quantum mechanics shows it is not possible to give 'a description of the motion of the corpuscles in our familiar space-time concepts' (Heisenberg, 1926a, p. 412). Significantly, Heisenberg argued that we are forced to abandon all hope of describing an electron in space and time 'unless one were to consider a space, whose structure [*Maßbestimmung*] differs substantially from the Euclidean one, as representing "ordinary" space' (Heisenberg, 1926a, pp. 412–13). In a letter to Pauli written on 28 October, Heisenberg set out his thoughts on the problem:

> [In quantum theory] it makes no sense, for example to speak about the position of a particle with definite velocity. However it does make sense if one does not consider

the velocity and the position too accurately. It is quite clear then that, macroscopically it is meaningful to speak about the position and the velocity of a body (all this is for you naturally nothing new). I have in all this a hope for a later solution of about the following kind, (but one should not say so too loudly): that space and time are actually only statistical concepts, as are temperature, pressure, etc. in a gas. I mean, space-like concepts and time-like concepts are meaningless for *one* particle, and that they become more and more meaningful the more particles are treated. I have tried often to come from this direction further, but as yet it does not want to succeed.

(Heisenberg to Pauli, 28 October 1926, Pauli, 1979, p. 350 [item 144])

Here we find Heisenberg's earliest anticipations of the uncertainty principle. However, unlike the version of the principle we find in the March 1927 paper, his view is underpinned by the possibility of a new conception of space and time. The suggestion that space and time are only statistical concepts in quantum theory had in fact been proposed by Rashevsky earlier that year (Rashevsky, 1926). To better understand the background to Heisenberg's formulation of the uncertainty relations, which appeared some four months later, it is worth emphasising that for Heisenberg the indeterminacy of the position and velocity of a particle was initially treated as a *problem*, and in this sense was in need of a solution. The emergence of the statistical interpretation of quantum mechanics through the work of Born and Pauli in the preceding months had already suggested that a statistical conception of space and time might be the right approach.

Reflecting further on the possibility that one might have to alter our basic concepts of space and time in the atomic world, Heisenberg speculated that the problem of interpreting quantum mechanics might well lie with the discontinuity of space-time itself. The concept of velocity in classical kinematics, Heisenberg explained in a letter to Pauli in November, is *by definition* dependent on the infinite divisibility of space and time. However, if the idea of an electron moving in a trajectory in space and time was no longer applicable in the atom, then perhaps one could no longer assume space and time were continuous. Writing to Pauli on 15 November 1926, Heisenberg raised the possibility of making sense of quantum mechanics through the introduction of a quantised space-time manifold:

If we hold space-time to be already discontinuous, it is then very satisfactory that it makes no sense to speak, for example, of the velocity v [of an electron] at a definite point x. For to define velocity one always needs at least two points, which [in a quantised space-time] can lie in a discontinuous relationship but *not* infinitely close. When we talk about position or velocity, we always need words that are obviously not defined at all in a discontinuous world.

(Heisenberg to Pauli, 15 November 1926, Pauli, 1979, pp. 354–6 [item 145])

This view had earlier been expressed by Max Born in 1919. In a letter to Pauli, Born had argued that the problems thus far encountered in quantum theory demand 'entirely new fundamental points of view', prompting him to suggest that: 'One is not allowed to carry over the concept of space-time as a four-dimensional continuum from the macroscopic world of experience into the atomistic world' (Born to Pauli, 23 December 1919, Pauli, 1979, p. 10 [item 4]). In 1922, Hans Reichenbach observed that 'under the influence of quantum theory', physicists had begun to entertain the possibility that it may be necessary 'to conceive of space as a discrete manifold' (Reichenbach, 1996, p. 275). A similar view is to be found in an unpublished manuscript written around 1926 in which Reichenbach argued that the theory of quantum mechanics recently developed by Heisenberg and Schrödinger demanded a 'shift in our conception of physical space'. Here he suggested that such a shift might take place 'by an additional number of dimensions or the introduction of a lattice-like manifold in which only the points of the lattice are occupied by matter'. However, Reichenbach noted that it was essential to preserve some kind of connection with 'a three-dimensional approximately Euclidean space' because 'for ordinary dimensions the old concept of space remains applicable' (Reichenbach, 1991, pp. 31–2).

As these remarks suggest, the possibility of a lattice-like structure of space appears to have attracted attention during this time and continued to feature prominently in Heisenberg's thinking. In his letter to Pauli in November, he stressed that all our usual concepts were bound up with the idea of a continuous space-time, but he was now firmly of the view that a continuous world was 'completely out of the question'. In a discontinuous world, it makes perfect sense that 'all the [usual] words that we apply to the description of an event' such as the position or velocity of a particle are rendered ambiguous (Heisenberg to Pauli, 23 November 1926, Pauli, 1979, pp. 357–60 [item 148]). Here, Heisenberg emphasised the limitations of our usual words and concepts. His speculations on the possibility of a statistical conception of space and time and the quantisation of the space-time manifold in October–November 1926 were attempts to radically transform the conceptual framework of quantum mechanics. This marks the continuation of a train of thought that appeared in Heisenberg's earlier deliberations on matrix mechanics, where he had portrayed the transition from 'classical visualisable geometry' to 'symbolic quantum geometry' as a key to the development of the new theory (Born, Heisenberg & Jordan, 1967, p. 322). As Kojevnikov points out, this dimension of Heisenberg's work has been largely neglected by historians of science, who have failed to recognise that 'the earliest incarnation of quantum mechanics – matrix mechanics – rested on the philosophical assumption that the classical notions of space and time become invalid inside the atom' (Kojevnikov, 2002, p. 420).

The problem posed by the limits of our classical language is again evident in a letter written to Pauli in February 1927. Taking issue with Jordan's paper on indeterminism and causality in quantum mechanics (Jordan, 1927a), Heisenberg posed the question: 'What for example does it mean to say the "probability that the electron is located at a definite point", if *the concept "position of the electron" is not properly defined*?' (Heisenberg to Pauli, 5 February 1927, Pauli, 1979, p. 374 [item 153], emphasis added). The problem of interpretation for him now properly rested with an analysis of how one should define the basic kinematic concepts of position and velocity in quantum theory. As he explained in the introduction to his 1927 paper:

> We have good reason to become suspicious every time uncritical use is made of the words "position" and "velocity". When one admits that discontinuities are somehow typical of processes that take place in small regions and in short times, then a contradiction between the concepts of "position" and "velocity" is quite plausible. If one considers, for example, the motion of a particle in one dimension, then in continuum theory one will be able to draw a worldline $x(t)$ for the track of the particle (more precisely for its centre of gravity), the tangent of which gives the velocity at every instant. In contrast, in a theory based on discontinuity there might be in the place of this curve a series of points of finite separation. In this case it is clearly meaningless to speak about a velocity at a position.
>
> *(Heisenberg, 1983, p. 63)*

Here, Heisenberg proposed that the 'contradictions evident up to now in the physical interpretation of quantum mechanics' would be resolved by carrying out 'a more precise analysis of the kinematical and mechanical concepts' such as position and velocity (Heisenberg, 1983, p. 63). This passage is of particular importance because it demonstrates two critical points. Firstly, it shows that at the time of writing the paper, Heisenberg was concerned first and foremost with the problem of the inadequacy of our usual concepts and words. There is no mention of the limits of measurement or of our knowledge, but instead the emphasis is placed squarely on the limits of language. To this end, Heisenberg proposed that we may 'arrive at a physical understanding' of quantum mechanics 'through a more precise analysis' of the concepts of position and velocity. Secondly, Heisenberg here stresses the importance of *discontinuity* in quantum mechanics, in understanding why it is not possible to specify *both* the position and the velocity of a particle. To this extent, the paper on the uncertainty relations should be read in the context of Heisenberg's earlier concerns, expressed in his correspondence with Pauli in October and November the previous year.

In his 1927 paper, Heisenberg again emphasised the discontinuity of quantum theory, as opposed to classical theory, but he now deemed 'a revision of space-time geometry at small distances' to be 'unnecessary' (Heisenberg, 1983, p. 62).

The significance of this observation becomes apparent only once we place it in the context of his earlier speculation concerning the possibility of transforming the conceptualisation of space and time in quantum mechanics. However, here, as before, Heisenberg maintained that 'a revision of kinematical and mechanical concepts is necessary' in quantum mechanics. But how does one redefine the basic concepts of position and velocity, while retaining the classical space-time geometry? Heisenberg's original thoughts on the problem of interpretation in quantum mechanics had revolved around the need to modify the fundamental concepts of space and time, but by late February 1927, he had turned his attention towards a different solution to the problem. The approach was now to invoke the verification theory of meaning that had gained currency in scientific and philosophical circles in the context of Einstein's special theory of relativity.

5.2 The introduction of the gamma-ray microscope

As we saw in Chapter 2, the idea that physical concepts are meaningful only if they correspond to measurable quantities had gained currency among physicists in the 1920s. In 1919, Wolfgang Pauli had criticised Hermann Weyl's attempts to develop a unified field theory of the electron on the grounds that the electric field strength inside an electron is unobservable in principle (Pauli, 1919, p. 749–50). Writing to Eddington in 1923, Pauli elaborated on this point, arguing that 'the field concept only has a *meaning* when we specify a reaction, which is in principle possible, by which we can *measure* the field strength at each point of space-time'. He concluded by saying that 'as soon as the reaction ceases to be specifiable or in principle executable the respective field concept is no longer defined' (Pauli to Eddington, 20 September 1923 Pauli, 1979, p. 117 [item 45], emphasis added). As noted earlier Pauli's view was largely inspired by Einstein's analysis of the concept of simultaneity in his 1905 paper on special relativity. In 1916, Einstein elaborated on this point explaining that the definition of simultaneity rests upon the possibility of determining the simultaneous occurrence of two events by means of experiment. 'As long as this requirement is not satisfied, I allow myself to be deceived ... when I imagine that I am able to attach a meaning to the statement of simultaneity' (Einstein, 1920, p. 26).

While the views of Pauli and Einstein did not constitute a precise philosophical system, this general point of view gained widespread acceptance among the positivists throughout the 1920s. The idea was given its clearest expression in the 1920s by Percy Bridgman in his book *The Logic of Modern Physics* (Bridgman, 1927; see also Bridgman, 1949). Bridgman used the term 'operationalism' to describe the new philosophical attitude to concepts that Einstein had introduced into physics. Under the section entitled 'Einstein's contribution in changing our attitude towards concepts', Bridgman argued that the meaning of the concept of length is determined when 'the operations by which length is measured are fixed: that is, the concept of length involves as much as and nothing more than the set of operations by which length is determined'. He went on to conclude that in physics 'we mean by any concept nothing more than a set of operations; *the concept is synonymous with the corresponding set of operations*' (Bridgman, 1927, p. 5, emphasis in original).

In this theory of meaning, which finds its clearest expression in Einstein's special theory of relativity, Heisenberg saw a judicious escape from the problems that had beset quantum theory from as early as 1924. The operational point of view rested on thought-experiments, most famously exemplified in Einstein's discussion of the definition of simultaneity. For Heisenberg, the meaning of concepts such as the position of an electron and velocity would acquire new meaning in precisely this way (Beller, 1988, p. 148). In a letter to Pauli dated 27 February 1927, which formed the draft for the paper that eventually went to print, Heisenberg juxtaposed two ways of framing the question: 'What does one understand by the words "position of the electron"? This question can be replaced according to the well-known model by the other: "How does one determine the position of the electron?" ' (Heisenberg to Pauli, 23 February 1927, Pauli 1979, p. 376 [item 154]). Heisenberg presented the same argument in the paper he submitted for publication in March:

> When one wants to be clear about what is to be understood by the words "position of the object" for example of the electron, then one must specify definite experiments with whose help one plans to measure the "position of the electron": *otherwise this word has no meaning*. There is no shortage of such experiments, which in principle allow one to determine "the position of the electron" with arbitrary accuracy. For example let one illuminate the electron and observe it under a microscope.
> *(Heisenberg, 1983, p. 64, emphasis added)*

Using the example of the imaginary gamma-ray microscope, which had earlier been suggested by Drude in Göttingen, Heisenberg explained that 'the highest attainable accuracy in the measurement of the position of the electron is governed by the wavelength of the light'. However, in such a scenario we can only measure the position of an electron at the expense of determining the precise momentum of the electron. This is because at 'the instant when position is determined – therefore, at the moment when the photon is scattered by the electron – the electron undergoes a discontinuous change in momentum'. The discontinuous exchange of energy between the incoming light quantum and the electron means that any measurement of the electron's position renders its momentum indeterminate 'up to magnitudes which correspond to that discontinuous change' (Heisenberg, 1983, p. 64). And to this extent, 'the more precisely the position is determined, the less precisely the momentum is known, and vice versa' (Heisenberg, 1983, p. 64).

We must not fall into the trap of reading Heisenberg's thought-experiment independently from the operational context in which it was proposed. To do this would be to interpret the gamma-ray microscope in purely *epistemic* terms, as a demonstration of the limits of our knowledge of the electron's position and momentum, rather than in *semantic* terms. The measurement of the position of an electron, in Heisenberg's paper, renders the concept of velocity *meaningless*. It is true that in later years Heisenberg sometimes referred to the gamma-ray micro-scope as demonstrating the limits on our knowledge of the electron's position and momentum, but in setting out the thought-experiment in 1927s Heisenberg was careful to point out that the instantaneous discontinuous change in the momentum of the electron at the point of observation means that we cannot ascribe a *well defined meaning* to the concept, when the position is determined. To this extent, Heisenberg claims that the very meaning of the concept of position is incompat-ible with the meaning of the concept of velocity in quantum mechanics, though both concepts can have a well-defined meaning when taken in isolation. The primary purpose of the thought-experiment was to secure the meaning of these concepts. One may of course argue – with some justification – that Heisenberg's attempt was misguided or poorly executed, but this should not blind us to the fact that in the introductory section of the 1927 paper, the thought-experiments were formulated to provide operational definitions for the key concepts in quantum mechanics including the path of the electron, velocity and the energy of the atom.

Heisenberg readily acknowledged the influence of Einstein's special theory of relativity on his approach to the definition of concepts in quantum mechanics. In his published writings and his private correspondence during this time, Heisenberg referred explicitly to the parallel between Einstein's redefinition of simultaneity in relativity, and the redefinition of concepts like position and velocity in quantum theory. This was made clear for the first time in a letter to Pauli on 9 March 1927, just weeks before the paper went to print. Heisenberg argued that Einstein's definition of simultaneity depends on the experimental possibility of measuring the time of two events (Heisenberg to Pauli, 9 March 1927, Pauli, 1979, pp. 383–4 [item 156]). In the published paper submitted a few weeks later, Heisenberg paid specific attention to the analogy:

> It is natural in this way to compare quantum theory with special relativity. According to relativity, the word "simultaneous" cannot be defined except through experiments in which the velocity of light enters in an essential way. If there existed a "sharper" definition of simultaneity, as, for example, signals that propagate infinitely fast, then relativity would be impossible ... We find a similar situation with the definition of the

concepts "position of an electron" and "velocity" in quantum theory. All experiments that we can use for the definition of these terms necessarily contain the uncertainty ($\Delta p.\Delta q \sim h$) ... even though they permit one to define exactly the concepts p and q taken in isolation.

(Heisenberg, 1983, p. 68)

This passage clearly demonstrates that Heisenberg had attempted to model his own approach to quantum mechanics on Einstein's operational viewpoint, which guided the development of special relativity. Using the imaginary gamma-ray microscope, Heisenberg was now able to specify definite experiments to determine the 'position of the electron'. In a short article entitled 'On the Fundamental Principle of Quantum Mechanics', published a month after the paper on the uncertainty relations, Heisenberg reiterated the view that while our classical intuitions suggest that it is 'always directly meaningful to speak about the "position" and "velocity" of a particle', in physics we must adhere to the operationlist standpoint that such 'words *only have a meaning* insofar as one can indicate the manner in which one can determine "position" and "velocity" by means of experimental measurement' (Heisenberg, 1927b, p. 83, emphasis added).

The crucial point for Heisenberg in devising the thought-experiment on the gamma-ray microscope was not primarily to determine the limits of precision of measurement, but rather to establish the precise meaning of concepts 'position' and 'velocity' in quantum mechanics through an operational analysis. To this extent, the gamma-ray microscope, as it was presented in his 1927 paper, was not intended primarily to demonstrate the limits of our knowledge of the electron's position and momentum, though it was frequently presented in this way in later years. It is a semantic, not an epistemic, concern that lies behind Heisenberg's original formulation of the thought-experiment in his paper on the uncertainty relations. This point has often escaped the notice of historians and philosophers in the subsequent presentations of Heisenberg's argument (Roychoudhuri, 1978; Favrholdt, 1994, p. 91; Röseberg, 1995, p. 108).

5.3 Bohr's analysis of the limits of measurement and the meaning of concepts

In March 1927, Bohr returned from Norway to find that Heisenberg had written the paper on the uncertainty relations. Bohr undoubtedly recognised that Heisenberg had made a decisive step in the search for a physical interpretation of quantum mechanics. But certain aspects of the paper troubled him. In the weeks immediately following Bohr's return, the uncertainty relations would become the subject of intense discussion and disagreement between Bohr and

Heisenberg. At the centre of the disagreement was the way Heisenberg had treated the limits of the accuracy of measurement in the gamma-ray microscope. This has been treated in detail by many historians of physics; however, it is worth carefully retracing the basic argument here.

Bohr's criticism of Heisenberg's treatment of the gamma-ray microscope rested on the view that the light ray impinging on the electron must be understood as both a wave and a particle – something Heisenberg had failed to do. Bohr insisted that one had to take into account the limits on the resolving power of the optical instrument due to the energy–momentum exchange between the incoming light quantum and the electron during measurement. For the purposes of the thought-experiment, Heisenberg had simply treated the electron as a point-particle that underwent a discontinuous transfer of energy and momentum in the interaction with a light quantum. In contrast, Bohr maintained that 'a discontinuous exchange of energy and momentum during observation' would not preclude one from 'ascribing accurate values to the space-time co-ordinates, as well as to the momentum–energy components *before* and *after* the process' (Bohr, 1928, p. 583). According to Bohr, while Heisenberg's uncertainty relation for the position and momentum of a particle was essentially correct, his account of the measurement scenario was not.

In his detailed conceptual analysis of the gamma-ray microscope, Bohr argued that when a beam of light impinges on an electron, the light must be conceptualised both as particle and wave. In the first instance it is necessary to think of the interaction between the electron and the light ray as an instance of Compton scattering, according to which a photon (or a light quantum) colliding with the electron will be deflected at a given angle. However, the light quantum interpretation alone cannot explain the diffraction of the light beam, nor can it account for the resolving power of the microscope. According to classical optics, the diffracted beam of light is not scattered in a definite direction, but rather spreads over a certain angle. In the gamma-ray microscope experiment, Bohr argued, the light ray 'spreads out' like a wave after interacting with the electron, and therefore it is impossible to determine precisely at what angle the photon is scattered (Murdoch, 1987, p. 49). The beam of light is then focused to a single point after having passed through the lens of a microscope. This is explicable only by means of the Rayleigh formula for the resolving power of an optical instrument, which in turn depends on the wave theory of light. For Bohr, it was not discontinuity of the interaction between the electron and the light quantum, but rather *the wave–particle duality of light* that prevented one from measuring the position and momentum of the electron with unlimited precision.

Writing to Pauli on 4 April, Heisenberg reported that in Copenhagen 'the thought-experiments are constantly discussed. I argue with Bohr over the extent to which the relation pq ~ h has its origin in the wave or the discontinuity aspect'. On the one hand Bohr had emphasised 'that in the γ-ray microscope the diffraction of the waves is essential', while on the other hand, Heisenberg had argued that 'the theory of light quanta and even the Geiger–Bothe experiment is essential' (Heisenberg to Pauli, 4 April 1927, Bohr, 1985, p. 17). After several weeks of discussion, Heisenberg conceded the point to Bohr. Replying to a letter from Dirac, Heisenberg explained that Bohr had convinced him that 'one cannot calculate the velocity' of the electron after measuring its position in the case of the gamma-ray microscope, not because of the discontinuous exchange of momentum, but 'because the direction of the dispersed light quantum after the collision is *not* known' (Heisenberg to Dirac, 27 April 1927, Bohr, 1985, p. 17). In his interview with Kuhn in 1963, Heisenberg recalled that, after discussions with Bohr, he had 'agreed about the interpretation of the gamma-ray microscope – that not only the discontinuities were important, but also this problem of the aperture and interference' (AHQP, 25 Feb 1963).

The Bohr–Heisenberg debate on the gamma-ray microscope did not mean a complete revision of the paper, which Heisenberg had already sent for publication, but simply a note appended at the end of the paper. There Heisenberg wrote: 'Bohr has pointed out that I have overlooked essential points in some of the discussions in this work'. In particular, he explained, 'in the use of an imaginary γ-ray microscope one must take into consideration the necessary divergence of the beam of radiation'. Consequently, 'in the observation of the electron position, the direction of the Compton recoil is only known with an uncertainty' (Heisenberg, 1983, p. 84). Writing to Pauli on 16 May, Heisenberg again admitted his earlier mistake:

> In respect to this work, Bohr has drawn my attention to the fact that in my paper yet another important point has been overlooked (Dirac also asked me about it later): With the γ-ray microscope one might first imagine that one determines the direction of the incident light quantum as well as the reflected light quantum. Then *after* the Compton effect both the position and the velocity are known very accurately (more accurately than $\Delta p \cdot \Delta q \sim h$). However, this cannot really be done, because of the diffraction of the light (wave theory!). To achieve an accuracy λ, the microscope must have an aperture of the order 1. Thus the relation $\Delta p \cdot \Delta q \sim h$ is of course obtained, but not in the manner I thought.
>
> *(Heisenberg to Pauli, 16 May 1927, Bohr, 1985, p. 19, emphasis in original)*

Bohr gave a full account of the measurement of the position of an electron using a gamma-ray microscope in his Como lecture in September 1927, which was published in slightly modified form in 1928. Here Bohr stressed the point: 'In using the optical instrument for determinations of position, it is necessary to remember that the formation of the image always requires a convergent beam of light'. Bohr employed Rayleigh's formula to calculate the resolving power of a microscope. He was therefore able to prove that the 'product of the least inaccuracies with which the positional co-ordinate and the component of the momentum in a definite direction can be ascertained is therefore given by [Heisenberg's uncertainty] formula' (Bohr, 1928, p. 583). In the years that followed, Heisenberg referred to Bohr's analysis of measurement in quantum theory, not to his own original formulation.

Although Heisenberg eventually accepted Bohr's point that the dispersion of the light beam was crucial for the uncertainty in the measurement of position and momentum, he was never convinced that it was necessary to use both wave and particle descriptions of light. As Heisenberg himself later explained, 'For explaining the γ-ray experiment, it was useful to play with both pictures ... but it was not absolutely essential. You could actually use both languages independently' (AHQP, 25 February 1963). Writing to Pauli on 4 April, Heisenberg reported that he and Bohr had continued to disagree about the extent to which the uncertainty relation 'has its foundations in the wave or discontinuity sides of quantum mechanics'. Yet, at this time, Heisenberg felt that there was little to be gained from such a discussion. As he explained, 'By overemphasizing one side or the other we can discuss much without anything new' (Heisenberg to Pauli, 4 April 1927, Pauli, 1979, pp. 390–3 [item 161]). Only months later did Heisenberg fully appreciate the philosophical point of view that lay behind Bohr's criticism of the gamma-ray microscope.

The disagreement between Bohr and Heisenberg was, on the surface at least, about the necessity of invoking the wave–particle duality of light to explain the limits of precision of measurement. But in order to more fully understand the nature of their disagreement we must examine more closely their respective philosophical positions. Once this is done we can get a clearer picture of the way in which Bohr's influenced Heisenberg's philosophical standpoint. Whereas for Heisenberg the very meaning of the words 'position of the electron' was made contingent upon the possibility of measuring the electron's position, Bohr held an altogether different view. As Max Jammer explains, the 'reduction of definability to measurability was unacceptable to Bohr ... the limitation of measurability confirms the limitation of definability but does not logically precede it' (Jammer, 1974, p. 69). This is clear from the Como lecture, where Bohr stated that the 'uncertainty which always affects' the measurement of the electron's position is 'essentially an outcome of the limited accuracy with which changes in energy and momentum can be *defined*' (emphasis added) in the interaction between the scattered light beam and the electron (Bohr, 1928, p. 583). In Bohr's account, the use of classical concepts such as position and momentum is presupposed in the description of the interaction.

The disagreement between Bohr and Heisenberg then should not be construed, as some scholars have done, as a dispute between the epistemic standpoint of Heisenberg and the semantic standpoint of Bohr, but rather as between *two different semantic conceptions*. Bohr's writings after 1927 contain numerous passages that confirm such a reading. In a paper published in 1929, Bohr wrote: 'It is therefore, an inevitable consequence of the limited applicability of the classical concepts that the results attainable by any measurement of atomic quantities are subject to an inherent limitation' (Bohr, 1987b, p. 95). Here we find perhaps the clearest statement of Bohr's view of the logical relationship between the limits of classical concepts and the limits of measurement. As John Hendry puts it, 'Bohr's analysis showed that ... an operational definition of the kinematic concepts needed was impossible' (Hendry, 1984, p. 126).

If we look deeper into the fundamental disagreement between Bohr and Heisenberg, we see that not only does it involve differing views concerning the definition of concepts, but it also rests on the different senses in which Bohr and Heisenberg were inclined to use the word 'definition' in their writings. This is a point that has not always been appreciated by scholars examining the Heisenberg–Bohr debate. Although Jammer and Hendry are certainly correct in arguing that Bohr implicitly rejected Heisenberg's operational reduction of definability to measurability, it is not sufficiently clear from their accounts that when Heisenberg spoke of the 'definition' of a concept, he intended something quite different from what Bohr did. Here it is worth quoting Edward MacKinnon's clarification of this point in his exposition of Bohr's position:

> Bohr did not believe that the meaning of such terms as "position" and "trajectory" could be determined by such operational definitions. These terms already have a determined meaning prior to and independent of any particular atomic experiments. The basic meaning of these terms is determined by ordinary language usage proper to classical physics. The critical problem requiring analysis accordingly is how such already meaningful classical concepts function in quantum contexts ... When Bohr speaks of "definition" ... he does not mean *stipulating* the meaning of some term or some concept. He refers rather to rules governing the usage, in quantum contexts, of a term whose classical meaning is already determined.
>
> *(MacKinnon, 1982, pp. 271, 274, emphasis added)*

MacKinnon's interpretation brings to light the quite different theories of meaning which lie behind the Heisenberg–Bohr dispute. When Heisenberg spoke of 'definition' in his paper, he did indeed mean stipulating the meaning of the concepts 'position' and 'velocity'. As MacKinnon explains, whereas Heisenberg had accepted a 'referential theory of meaning', Bohr took the view that 'meaning came from its usage in ordinary language and the idealization of that usage in classical physics' (MacKinnon, 1984, p. 177). Thus, when Bohr says the concept of position is not 'well-defined', he does not mean we do not understand the meaning of the word 'position', but rather that the concept cannot strictly speaking be applied to a particle. It was in this context that Bohr argued in the Como paper, 'an unambiguous definition' of the concept of velocity 'is excluded by the quantum postulate' (Bohr, 1928, p. 583).

In a letter to Bohr, written in June 1927, Heisenberg explained that only after Pauli's visit had he fully grasped the importance of what Bohr was trying to say. 'Now I have understood much better that it really matters very much to put the concepts in the order of precedence which you want and not how I put it in my paper; and I also see very well that in this way it has certainly become much more beautiful' (Heisenberg to Bohr, 18 June 1927, Mehra & Rechenberg, 2000, p. 186). In the discussions that took place at the Como conference in September 1927, Heisenberg explained: 'The physical interpretation of the uncertainty relation $\Delta p \, \Delta q \sim h$ and its relationship with the general points of view raised by Bohr have been made entirely clear for the first time through the investigations of Bohr' (Bohr, 1985, p. 141). It should be noted that Heisenberg's remarks at the Como conference show that in September 1927 he was still attempting to draw a parallel between Einstein's definition of simultaneity in the special theory of relativity and the redefining of concepts in quantum mechanics along operationalist lines. However, as we shall see in the section below, by 1929 Heisenberg appears to have been won over to Bohr's philosophical point of view. My claim here is not that there is a precise moment when Heisenberg changed his position, but that in the months that followed his publication of the uncertainty principle paper his thinking underwent a gradual but significant shift. Indeed, we should not forget that during this time Heisenberg's thinking on a range of other issues such as observability and wave–particle duality were still in a state of transition, and had not yet fully crystallised.

5.4 The turning point in Heisenberg's philosophy

Before examining in more detail Heisenberg's conversion to Bohr's philosophical point of view, it is worth briefly recapping the train of thought that had guided Heisenberg in the months immediately preceding his clash with Bohr in Copenhagen. Prior to his paper in 1927, Heisenberg had been preoccupied with the development of a new conceptual framework for kinematics, one that involved a new understanding of space and time at the sub-atomic level. The 1927 paper should therefore be read in the context of these concerns. At this time, Heisenberg was concerned with the fundamental question: 'What does one understand by the words "position of the electron"?' (Heisenberg to Pauli,

23 February 1927, Pauli, 1979, p. 376 [item 154]). In October–November 1926, Heisenberg had pursued the possibility of a 'discontinuous' space-time, or a statistical interpretation of space-time concepts. However, by February 1927, he approached the question from a different point of view. Heisenberg hoped that the operational program Einstein had successfully implemented into the theory of relativity would again prove its worth in quantum mechanics.

Bohr, on the other hand, had never approached the problem in such a way. For Bohr, the tension between classical concepts and the quantum world was not something that could be explained away through a deeper level of physical or philosophical analysis, but something we had to accept and take for granted as a point of departure. The limited applicability of classical concepts was, for him, a situation that could not be overcome through an operational analysis of the meaning of concepts. Note the contrast between the two: in the paper on the uncertainty relations, Heisenberg had stated that 'no interpretation of quantum mechanics is possible which uses ordinary kinematic and mechanical concepts' (Heisenberg, 1983, p. 62), whereas in the Como paper Bohr took the view that any 'interpretation of the experiments rests essentially upon the classical concepts' (Bohr, 1928, p. 580).

It was only some months after the publication of the paper on the uncertainty relations that Heisenberg came to appreciate the philosophical point of view behind Bohr's criticisms of the gamma-ray microscope. In an interview with Kuhn in 1963, Heisenberg recalled that he and Bohr had initially approached the problem of defining concepts in quite different ways. As he explained, in the uncertainty principle paper, he had 'not yet completely' abandoned his hope of replacing the concepts of classical physics with quantum concepts. While we should be mindful of the fact that Heisenberg's later recollections are not always trustworthy, it is worth quoting at length from his interview with Kuhn in 1963, in which Heisenberg spells out the shift that occurred in his thinking in his discussions in Copenhagen:

> Well, what is clearly in that paper is my own misunderstanding that one could not use the words "position" and "velocity" in the same manner as one had done before. So these words ceased to get hold of the phenomena. Now the general attitude is that *one still has to keep the words in spite of the fact that they have these limitations* – that was Bohr's point. Bohr would insist, "Well, in spite of your Uncertainty Principle you have got to use the words 'position' and 'velocity', just because you haven't got anything else." Well this was Bohr's side of the picture which came out during the discussions which probably in the paper were not so clear to me as they were a few months later. It was just in this stage of the development that one gradually became accustomed to the idea that we never really can get out this atmosphere of despair and hopelessness because we never have the words by which we can really do the thing.
>
> *(AHQP, 27 February 1963, emphasis added)*

Earlier in the interview with Kuhn, Heisenberg had made a similar point:

> it is also true to say that just by the discussions with Bohr I learned that *the thing which I in some way attempted could not be done*. That is, one cannot go entirely away from the old words because one has to talk about something. I saw that very clearly in this gamma-ray microscope. In the gamma ray microscope you must say what happens in your microscope ... So I could realize that I could not avoid using these weak terms which we always have used for many years in order to describe what I see ... I realized in discussing these problems with Bohr, how desperate the situation is. On the one hand we know that our concepts don't work, and on the other hand, we have nothing except the concepts with which we could talk about what we see.
>
> *(AHQP, 27 February 1963, emphasis added)*

Only after discussions with Bohr did Heisenberg realise that the concepts of classical physics – position and momentum – are indispensable for a *description* of what occurs during the interaction between the electron and a light quantum in the gamma-ray microscope. Under Bohr's critical influence, Heisenberg resigned himself to the fact that the situation, which had troubled him in the months prior to the publication of the uncertainty principle paper, would now just have to be endured. As Heisenberg himself would put it, 'I think that this tension you just have to take; you can't avoid it. That was perhaps the strongest experience of these months – that gradually I saw that one will always have to live under this tension. You could never hope to avoid this tension' (AHQP, 27 February 1963). By late 1927, or perhaps early 1928, Heisenberg had effectively abandoned his earlier hope, expressed in the paper on the uncertainty relations, that one could find new empirical concepts for quantum mechanics.

We can find strong evidence for this shift in Heisenberg's thought by contrasting the philosophical position articulated in his 1927 paper with that of his 1929 Chicago lectures, which were published as *The Physical Principles of the Quantum Theory*. This shift in Heisenberg's thought is evident in three closely related but distinct ways. The first of these is his explicit disavowal of the operational point of view. Reflecting on his earlier attempts to 'avoid the contradictions' in quantum mechanics, Heisenberg restated the operational standpoint that had guided his efforts in 1927, namely that 'it seems necessary to demand that no concept enter a theory which has not been experimentally verified at least to the same degree of accuracy as the experiments to be explained by the theory'. But now he admitted that: 'Unfortunately it is quite impossible to fulfil this requirement, since the commonest ideas and words would often be excluded' (Heisenberg, 1930, p. 1). By 1929, Heisenberg was not defining concepts, but the limits of their applicability. This conceptual shift coincides with the shift in Heisenberg's use of terms. Whereas in 1927 he was more inclined to use the term *Ungenuaigheit* (imprecision) in connection with measurement, after 1930 he was far more inclined to employ the term *Unbestimmtheit* (indeterminacy) in connection with the use of classical concepts.[3]

The second way in which this shift in Heisenberg's thought manifested itself concerns the analogy he had drawn between the theory of relativity and quantum mechanics. In both the 1927 paper and his letter to Pauli on 9 March of that year, quoted earlier, Heisenberg had explicitly stressed the importance of using an operational definition of concepts in both theories. However, by 1929, the analogy between quantum mechanics and relativity in this context had disappeared altogether from Heisenberg's writings. Indeed, it is worth noting that in the 1929 lectures Heisenberg refrained from arguing, as he had in 1927, that Einstein's special theory of relativity was grounded in an operational definition of simultaneity. More importantly, he now *contrasted* the theory of relativity with quantum mechanics. In an interview conducted by David Peat in 1975, Heisenberg took issue with the suggestion that 'quantum mechanics has modified language' in much the same way as occurred with relativity:

> In the case of relativity theory, I would agree that physicists have simply modified their language; for instance, they would use the word *simultaneous* now with respect to certain coordinate systems. In this way they can adapt their language to the mathematical scheme. But in quantum theory this has not happened. Physicists have never really tried to adapt their language, though there have been some theoretical attempts. But it was found that if we wanted to adapt the language to the quantum theoretical mathematical scheme, we would have to change even our Aristotelian logic. That is so disagreeable that nobody wants to do it; it is better to use the words in their limited senses, and when we must go into the details, we just withdraw into the mathematical scheme.
>
> *(Peat & Buckley, 1996, p. 8, emphasis in original)*

Commenting on the later attempts to develop a new language suited to quantum mechanics, Heisenberg stressed that later it was realised that any attempt to modify our language would require the abandonment of standard Aristotelian two-value logic. While some attempts were later made in this direction (Reichenbach, 1944; Weizsäcker, 1955), in Heisenberg's view they did not really resolve the fundamental ontological problems of quantum mechanics. Mara Beller has argued that Heisenberg's willingness to speak of the adjustment

---

[3] Max Jammer notes that in his 1927 paper Heisenberg uses the term *Ungenauigkeit* (inexactness or imprecision) and *Genauigkeit* thirty times, whereas the term *Unbestimmtheit* (indeterminacy) is used only twice and *Unsicherheit* (uncertainty) is used three times (Jammer, 1974, p. 61). After 1930, Heisenberg used the term *Unbestimmtheit* almost exclusively (Mehra & Rechenberg, 2001, p. 811).

to new concepts of space and time in relativity shows that he did not hold steadfast to the indispensability of classical concepts in quantum mechanics (Beller, 1999, pp. 182, 197–9). Yet in his later writings and interviews, Heisenberg was careful to distinguish these as two quite different cases. In an interview with Kuhn in 1963, Heisenberg had explained precisely this point – that in the case of the theory of relativity, the concepts of space and time had acquired new meaning, whereas in the case of quantum mechanics, physicists had never adjusted classical language to suit the mathematical scheme (AHQP, 27 February 1963).

A further comment on Heisenberg's doctrine of classical concepts seems in order here. It might be argued that Heisenberg was inclined to abandon the indispensability of classical concepts on occasions. In the passage quoted above, for example, Heisenberg suggests that 'when we must go into the details, we just withdraw into the mathematical scheme'. Heisenberg certainly advocated using the mathematical formalism of quantum mechanics to describe the state of an isolated quantum system. However, he remained committed to the view that the *results of measurements* must be interpreted along classical lines and here we simply *cannot go into the details*. I will elaborate on this point further in Chapter 7. That one could not find a new language for describing our experimental results, Heisenberg conceded, 'was not so clear at the time' he was writing the uncertainty paper, but after discussions with Bohr, it seemed 'the only sensible thing to do was to use the old words and always remember their limitations' (AHQP, 27 February 1963). Indeed, this turns out to be a crucial point in understanding Heisenberg's epistemological position as it developed from 1930 onwards. As he was to put it: 'the Copenhagen interpretation of quantum theory starts with a paradox'. On the one hand we must 'describe our experience in the terms of classical physics' while on the other hand we know 'that these concepts do not fit nature accurately' (Heisenberg, 1958d, p. 55). This paradox, which Heisenberg did not fully appreciate until some time *after* the publication of the paper on the uncertainty relations in 1927, was the enduring legacy of his discussions with Bohr in Copenhagen.

The third way in which we can discern a shift in Heisenberg's thought after 1927 can be found in the *role* he accorded the gamma-ray microscope thought-experiment in his 1929 lectures. In his 1927 paper, Heisenberg had initially proposed the thought-experiment as a means of establishing an operational definition of concept of 'position'. But in the 1929 lectures he simply presented the gamma-ray microscope as an illustration of the inherent limits of the accuracy in measuring the position and momentum of an electron. It has gone largely unnoticed that in Heisenberg's 1929 Chicago lectures, the gamma-ray microscope no longer serves the purpose for which it was originally intended – namely, to provide an operational definition of the concepts position and velocity. Nowhere there do we find an attempt to define the meaning of concepts such as 'position' and 'velocity'. There, the gamma-ray microscope thought-experiment merely serves to *illustrate* the indeterminacy in the position and momentum of an electron. This shift in Heisenberg's understanding of the role of the gamma-ray microscope has been largely overlooked. Melanie Frappier, for example, argues: 'Despite the replacement of the derivation of the microscope thought experiment by the one using Bohr's complementarity principle, Heisenberg's general 1929 argument in favour of quantum mechanics preserves the line of argument followed in the 1927 uncertainty paper' (Frappier, 2004, p. 85). While this has remained the prevailing view in much of the literature, on closer inspection it is evident that Heisenberg's analysis of the indeterminacy principle in his 1929 Chicago lectures is based on a quite different philosophical position to the one we find in his 1927 paper.

In later years, Heisenberg referred to Bohr's analysis of the limits of accuracy in measuring the electron's position and momentum to some extent as 'trivial' (Heisenberg, 1977, pp. 5–6). If one assumes from the outset that the electron and the light quantum do not each *possess* a well-defined position and velocity, an inevitable uncertainty surrounds the *measurement* of position and momentum of the electron. In 1930, Heisenberg asked his student, Carl Friedrich von Weizsäcker, to perform a detailed theoretical analysis of the limits of accuracy for measuring the position and momentum of an electron with a gamma-ray microscope using a quantum field theory of radiation. Weizsäcker's analysis, which he published in 1931, and which was the subject of his doctoral dissertation, constitutes the final chapter in the history of the gamma-ray microscope thought-experiment (Weizsäcker, 1931). Weizsäcker's paper forms the natural extension of Bohr's work, in confirming that that the indeterminacy of the position and momentum of an electron using a gamma-ray microscope could be calculated through a quantum electrodynamic analysis (Frappier, 2004, p. 108–9).

Heisenberg did sometimes express himself as if he held a classical disturbance theory of measurement, according to which it is only the perturbation brought about by the act of observation that precludes us from knowing the precise position and momentum of the electron. Certain passages from the 1929 lectures invite such a reading. As Heisenberg explained there, the 'observation of the position will alter the particle's momentum by an unknown and undeterminable amount such that *after* carrying out the experiment our knowledge of the electronic motion is restricted by the uncertainty relation' (Heisenberg, 1930, p. 20). This way of putting things seems to suggest that *before* the measurement has taken place the particle does indeed have a well-defined position and momentum, even if we cannot know them. It was for this very reason that in his 1938 Warsaw paper, Bohr warned against using expressions like 'disturbance through measurement' which tended to unnecessarily confuse the issue (Bohr, 1996a). In an effort to clarify the misunderstandings which had arisen regarding the meaning of the uncertainty relations, Moritz Schlick argued that contrary to some popular accounts, the electron does not possess a definite position and momentum, and we simply 'have to renounce the desire to know them both exactly' (Schlick, 1979c, p. 485). In a similar vein, Ernst Cassirer warned: 'We cannot say the electron, at a given time, "is" at a certain location, nor that it "possesses" a sharply defined velocity, when this "possession" exists for the electron itself and not for physical knowledge' (Cassirer, 1956, p. 179).

Though on occasions Heisenberg invoked the classical imagery of a disturbance through measurement in his writings and lectures, it should be clear enough from the way his thinking unfolded after his paper on matrix mechanics in 1925, that he never seriously entertained the idea that an electron has a well-defined position and momentum. As he explained in the 1929 Chicago lectures: 'This indeterminateness is to be considered as an essential characteristic of the electron' (Heisenberg, 1930, p. 14). In his more careful moments, Heisenberg explicity rejected a purely epistemic interpretation of the uncertainty relations. In his lecture on 'The Role of the Indeterminacy Relations in Modern Physics' in December 1930, Heisenberg insisted that 'the indeterminacy relations hence should not simply be conceived of as the impossibility of precisely *knowing* or *measuring* the position and velocity [of an electron]; the indeterminacy relations signify that an *application of the words* "position, velocity" loses any reasonable meaning beyond specified limits' (Heisenberg, 1931b, p. 367, emphasis added).

It is worth emphasising that the positivism which appears to underpin two of his most important papers on quantum mechanics – the 1925 paper which formed the basis for matrix theory of kinematics and mechanics and the 1927 paper in which he formulated the uncertainty relations – are not at all representative of Heisenberg's later philosophical viewpoint. By the 1930s, Heisenberg's emphasis on observability and operationalism, which reveals a decidedly 'positivist' inclination in his early thought, had all but disappeared. An appreciation of this important point reinforces the need to attend to the historicity of Heisenberg's thinking. Though both papers are considered seminal contributions in the development of quantum mechanics, the philosophical views contained therein do not, as it turns out, reflect Heisenberg's mature views. Rather, they were forged during a transitional stage in his thought, while his ideas were still taking shape.

Though the discussion with Bohr in Copenhagen left a lasting impression on Heisenberg, there would remain important differences between the two thinkers. We have already seen that they disagreed over the meaning of wave–particle duality. But Heisenberg's departure from Bohr is also evident in his interpretation of Bohr's general concept of complementarity, which forms the subject of the next chapter. While Heisenberg incorporated some version of complementarity into his own interpretation of quantum mechanics, as will become apparent, this would take a decidedly different form than the one we find in Bohr's philosophical work.

6

Heisenberg and Bohr: divergent viewpoints of complementarity

In 1927, Niels Bohr announced a general framework for understanding quantum mechanics, which he termed complementarity. Though the meaning of complementarity has been the subject of much debate and disagreement, it is widely thought to constitute Bohr's most important contribution to the interpretation of quantum mechanics, as many of Bohr's contemporaries acknowledged (Pauli, 1994). Jordan described it as 'the most significant result for philosophy that crystallized out of modern physics' (Jordan, 1944, p. 131). In 1933, Pauli went so far as to suggest that modern quantum theory might be referred to as the 'Theory of Complementarity' (Pauli, 1980, p. 7). Yet, Bohr's contemporaries, many of whom saw themselves as adherents of the complementarity viewpoint, often failed to grasp Bohr's original meaning. Indeed, Don Howard has recently argued: 'Disentangling Bohr's views from those who claimed to speak on his behalf' remains important for 'understanding what complementarity really involves' (Howard, 2004, p. 670). This chapter contributes to just this task by examining the way in which Heisenberg reinterpreted Bohr's view of complementarity.

Although Heisenberg publicly defended Bohr's view of complementarity throughout the late 1920s and 1930s, his understanding of Bohr's original formulation of complementarity turns out to diverge considerably from Bohr's. In this chapter, I concentrate on their respective interpretations of the complementarity thesis for space-time and causal descriptions as well as the complementarity of quantities like position and momentum in mutually exclusive experimental arrangements. It has often passed unnoticed that in the introduction to the Como paper, in which he first publicly announced his view of complementarity, Bohr had intended to deal with the problem of stationary states, and he did not invoke an argument for the use of mutually exclusive experimental arrangements characteristic of his later versions of the complementarity thesis. While not wishing to overstate the significance of the different illustrations of Bohr's concept of complementarity, it is worth pointing out that scholars have tended to ignore the original formulation of complementarity in the introduction to the Como lecture, instead preferring to focus their analysis on the complementarity of mutually exclusive experimental arrangements which Bohr emphasised throughout the 1930s. This tendency is also evident even in the works of Bohr's contemporaries, notably Pauli and Jordan (Jordan, 1936, pp. 114–30; Pauli, 1980, p. 7).

In the 1927 Como lecture, Bohr employed the term 'causal description' to refer to the conservation of energy, while a space-time description referred to pinpointing the electron's position in space at a given time. In attempting to give a reconstruction of Bohr's basic argument in the late 1920s and early 1930s, Heisenberg presented a different account of Bohr's notion of complementarity, by interpreting the causal description of a system as a description of the evolution of the ψ-function. According to Heisenberg, the idea of complementarity does not express a relationship between classical concepts, but rather refers to the necessity of invoking two modes of description in quantum mechanics – the experimental (expressed in classical concepts) and the formal (expressed through the Schrödinger equation).

In his discussion of the case of the kinematic-dynamic state of a free particle, Heisenberg agreed with Bohr that the experimental conditions for the measurement of the position and momentum of a particle were mutually exclusive, but disagreed with Bohr that such experimental arrangements serve to define the very conditions for the *unambiguous* use of the concept of position or momentum. In fact for Heisenberg, complementarity highlighted the *ambiguity* inherent in the use of classical concepts in quantum mechanics. Bohr and Heisenberg also disagreed about certain aspects of the analysis of the dividing line between object and measuring instrument, which was central to Bohr's analysis of measurement. By bringing to light the divergent views of Bohr and Heisenberg regarding complementarity, we can gain a better insight into their different philosophical attitudes to the interpretation of quantum mechanics more generally.

6.1 Bohr's complementarity of space-time and causal descriptions

On 16 September 1927, Bohr presented his celebrated paper on 'The Quantum Postulate and the Recent Developments of Atomic Theory', at the Como conference (Bohr, 1985, pp. 113–36). It was here that he first announced publicly his view that quantum mechanics demanded a new philosophical standpoint, which he termed 'complementarity', though he had privately discussed this idea with Heisenberg, Pauli and Klein in Copenhagen as early as February 1927. The central departure from classical physics, as Bohr now saw it, was the separation of the space-time description in quantum mechanics from what he termed the causal description. In the introduction to his paper he characterised these two modes of description as *complementary* but *mutually exclusive*:

> On one hand, the definition of a state of a physical system, as ordinarily understood, claims the elimination of all external disturbances. But in that case, according to the quantum postulate, any observation will be impossible, and above all, the concepts of space and time lose their immediate sense. On the other hand, if in order to make observation possible we posit certain interactions with suitable agencies of measurement, not belonging to the system, an unambiguous definition of the state of the system is naturally no longer possible, and there can be no question of causality in the ordinary sense of the word. The very nature of the quantum theory thus forces us to regard the space-time co-ordination and the claim of causality, the union of which characterizes the classical theories, as *complementary* but *exclusive* features of the description, symbolizing the idealization of observation and definition respectively.
>
> *(Bohr, 1928, p. 580, emphasis added)*

Referring to this passage, Max Jammer has pointed out, 'This statement in which the term "complementarity" appears for the first time ... contained the essence of what later became known as the "Copenhagen interpretation" of quantum mechanics' (Jammer, 1966, pp. 351–2). However, there is considerable disagreement today among scholars over the philosophical meaning of Bohr's notion of complementarity (Faye & Folse, 1994, pp. xxvi–xxvii; Held, 1994). Much of the confusion surrounding Bohr's views stems from the ambiguous wording of many of the critical passages in his writings. Indeed the difficulties of interpreting Bohr's words were keenly felt by Bohr's contemporaries, even those who worked closely alongside him. As we shall see, Heisenberg's own understanding of this crucial passage turns out to be quite different from Bohr's intended meaning.

It is first necessary to examine Bohr's idea that space-time and causal descriptions are complementary. A space-time description for Bohr meant determining the position of an electron in space at a given time. This is unproblematic. But Bohr's understanding of what is meant by the 'causal description' requires further elucidation. For Bohr, causality in physics was virtually synonymous with the conservation of momentum and energy. As he put it: 'The fact that such a closed system is associated with a particular energy value may be considered as an immediate expression for the claim of causality contained in the theorem of the conservation of energy' (Bohr, 1928, p. 587). In a letter to Schrödinger on 23 May 1928, Bohr again made it clear that his argument rested on 'the complementary nature of the space-time coordinates and the *conservation laws*' (Bohr, 1985, p. 48, emphasis added). Here, Bohr interpreted the causal description in physics through the conservation of energy and momentum – a view he had expressed as early as 1924 in his joint paper with Kramers and Slater on the virtual oscillator model of the atom. There, Bohr, Kramers and Slater had argued that 'the causal connexion between the transitions in distant atoms' involves the 'direct application of the principles of conservation of energy and momentum, so characteristic of classical theories' (Bohr, Kramers & Slater, 1924, p. 791).

The critical point in Bohr's complementary description is that, in defining the 'state of a physical system' as entailing 'the elimination of external disturbances', he was not referring to the state of a particle described by ψ-function in configuration space, but rather to the well-defined *energy* of a *stationary state*. In his paper, Bohr explains that 'the conception of stationary states involves strictly speaking the exclusion of all interactions with individuals not belonging to the system'. This fact, Bohr tells us, 'forms the necessary condition for an *unambiguous definition of the energy of the atom*' (Bohr, 1928, p. 587, emphasis added). For Bohr, the very concept of the stationary state, with a well-defined energy, is incompatible with a description of the trajectory of the electron in the atom. Writing to Schrödinger on 23 May 1928, Bohr emphasised that his concept of complementarity was intended to account for stationary states:

> There remains always – as stated in the article – an absolute exclusion between the application of the concept of stationary states and the tracking of the behaviour of an individual particle in the atom. This exclusion provides in my opinion a particularly striking example of the general complementary nature of the description. As I have tried to show in my article, a quite definite meaning can be ascribed to the concept of stationary states as well as to the discrete energy values within their domain of applicability.
>
> *(Bohr, 1985, p. 49)*

It is significant here to note that this form of complementarity is *not* dependent on mutually exclusive experimental arrangements. Rather, it sets up a complementary relationship between the stationary state of the atom, characterised by its energy, and the experimental determination of the electron's position in space at a given time. In his Faraday Lecture delivered before the Fellows of the Chemical Society in May 1930, Bohr again emphasised: 'the idea of stationary states stands in a mutually exclusive relationship to the applicability of space-time pictures'. Here, Bohr acknowledged that the application of the concept of energy for the stationary state is somewhat paradoxical, when we consider 'that the very idea of motion, on which the classical definition of kinetic energy rests, has become ambiguous in the field of atomic constitution' (Bohr, 1932, pp. 375–6).

It is interesting to note that although Bohr discusses the concept of a stationary state in the passage quote above, a free particle may also serve as illustration of complementarity in much the same way.[1] From this point of view, a particle (say an electron) may also have a well-defined energy and momentum, provided it is not interacting with a measuring instrument. In this case, the electron would not be represented as a localised wave packet, but by a single harmonic plane wave extended infinitely throughout space in one dimension. However, to my knowledge, Bohr never explicitly presented the complementarity for a free particle in this way. On the occasions when he did consider the case of complementarity involving the position and momentum of a particle in quantum mechanics, he did so from the perspective of mutually exclusive experimental arrangements. We will return to this illustration of complementarity in quantum mechanics later in the chapter.

Much of the scholarly attention on complementarity has focused on Bohr's idea of the necessity of mutually exclusive experimental arrangements for a determination of the position and momentum of a particle, which he first articulated in the concluding sections of the Como paper in 1927. However, I want to first focus on the way in which Heisenberg interpreted the version of complementarity Bohr presented in the introduction to the Como paper concerning stationary states. At first glance, Heisenberg's description of the state of affairs in quantum theory appears to be a mere paraphrasing of Bohr's introduction. On closer inspection, however, we can discern a fundamental difference between the two.

6.2 The Heisenberg interpretation

Heisenberg's first reference to Bohr's complementarity argument can be found in a lecture Heisenberg presented at Leipzig in 1928. Echoing Bohr's words, Heisenberg declared that 'the causal description of a system is complementary to the space-time description, because, in order to obtain a space-time description, one must observe it, and this observation disturbs the system. If the system is disturbed, we cannot follow anymore its causal connection in a pure manner' (Heisenberg, 1984a, pp. 26–7). Similarly, in his 1929 Chicago lectures, published as *The Physical Principles of the Quantum Theory* in 1930, Heisenberg explained that causality only holds for 'isolated systems' and such systems

---

[1] I would like to thank one of the anonymous reviewers of a paper I submitted to the journal *Studies in the History and Philosophy of Modern Physics* for pointing this possibility out to me.

'cannot be observed'. He then gave the following exposition of what he presented as Bohr's concept of complementarity:

> As is clear from what has been said, the resolution of the paradoxes of atomic physics can be accomplished only by further renunciation of old and cherished ideas. Most important of these is the idea that natural phenomena obey exact laws – *the principle of causality* ... Second among the requirements traditionally imposed on a physical theory is that it must explain all phenomena as relations between objects existing *in space and time* ... Bohr has pointed out that it is therefore impossible to demand that both requirements be fulfilled [simultaneously] in quantum theory. They represent *complementary* and *mutually exclusive* aspects of atomic phenomena. This situation is clearly reflected in the theory which has been developed. There exists a body of exact mathematical laws but these cannot be interpreted as expressing simple relationships between objects existing in space and time.
>
> *(Heisenberg, 1930, pp. 62–4, emphasis added)*

It is clear from the passage above that Heisenberg recognised Bohr's Como paper on complementarity, which was published (with minor changes) in the English journal *Nature* and the German journal *Naturwissenschaften* in April 1928, as an important contribution to the philosophical understanding of quantum mechanics. As Heisenberg explained: 'It is only after attempting to fit this fundamental complementarity of space-time description and causality into one's conceptual scheme that one is in a position to judge the degree of consistency of the methods of quantum theory' (Heisenberg, 1930, p. 65). Remarks such as these are commonly read as indicating Heisenberg's acceptance of Bohr's viewpoint of complementarity. In his study of Heisenberg's philosophy of quantum mechanics, Patrick Heelan writes somewhat disparagingly of what he terms Heisenberg's 'capitulation to the philosophy of complementarity' (Heelan, 1965, p. 44). In similar vein, Mehra and Rechenberg conclude that Heisenberg's 1929 lectures represent 'nothing but variations on the theme of complementarity as the leading concept' (Mehra & Rechenberg, 2000, p. 331). Heisenberg himself gives this impression in the preface to *The Physical Principles of the Quantum Theory*, in writing that his work 'contains nothing that is not to be found in previous publications, particularly the investigations of Bohr' (Heisenberg, 1930, p. x).

However, a careful reading shows that Heisenberg's interpretation of complementarity differed from Bohr's. This divergence centres on the different ways in which they defined 'causality'. Citing Bohr's paper in a section entitled 'Bohr's concept of complementarity', Heisenberg re-interpreted the causal description in quantum mechanics as a description in terms of the deterministic evolution of the Schrödinger wave function. This is evident in his diagrammatic representation of complementarity in his 1929 lectures, in which we may either describe phenomena in terms of space and time, but within the limits specified by the uncertainty principle, or alternatively we can give a causal description using the mathematical laws of quantum mechanics, thereby rendering a space-time description of the phenomena impossible. In this reconstruction of Bohr's ideas, Heisenberg departs from the view that the uncertainty principle presents a special case of complementarity, and instead refers to it as expressing a limitation to the space-time description of phenomena, which is complementary to the causal description.

In his lecture 'The Causal Law and Quantum Mechanics' in September in 1930, Heisenberg argued that it is possible to speak of the breakdown of causality in quantum mechanics because 'the future behaviour' of any system 'can be predicted [only] inaccurately, i.e., only statistically'. However, Heisenberg maintained that if one defines the concept of causality in a more 'restricted' way, as referring to the deterministic evolution of the ψ-function, it is possible to say that causality is retained in quantum theory (Heisenberg, 1931a, p. 177). He was well aware that this marked a departure from the view he had expressed in the concluding section of his 1927 paper on the uncertainty relations. There, Heisenberg had argued that quantum mechanics breaks with a causal description because it is not possible to ascertain the precise position and momentum of a particle at a given time, thereby precluding the possibility of calculating the motion of the particle at any subsequent time (Heisenberg, 1983, pp. 83–4). In a lecture in December 1930, Heisenberg explained that 'the physical behaviour of the system in general can only be statistically predicted from the Schrödinger function' but he now added, 'the function itself can also be calculated precisely even for later times, as far as the system remains undisturbed' (Heisenberg, 1931b, p. 370). Here, Heisenberg conceded that it 'is purely a question of taste' whether one wishes to regard 'this kind of regularity as causal or not' (Heisenberg, 1931b, p. 370). In the introduction to Heisenberg's *Physics and Philosophy*, Northrop draws attention to the distinction between Heisenberg's use of the terms 'causality' and 'determinism' in this context (Northrop, 1958, pp. 18–20).

Heisenberg's redefinition of causality had been suggested by a number of other physicists during this time. In Max Born's second paper on electron scattering collisions in 1926, he observed, 'the probability [function] itself propagates in harmony with the causal law' which takes the form of the differential wave equation (Born, 1968, p. 208). Similarly, in discussions at the Solvay conference in October 1927, Dirac had argued, quantum mechanics 'describes the state of the world at any instant by a wave function ψ which normally develops according to a causal law so that its initial value determines its value at any later instant' (Bohr, 1985, p. 104). In *Physics and Philosophy*, Heisenberg reinterpreted Bohr's view of complementarity in precisely this way:

> Bohr uses the concept of 'complementarity' at several places in the interpretation of
> quantum theory ... The space-time description of the atomic events is complementary
> to their deterministic description. The probability function obeys an equation of
> motion as did the co-ordinates in Newtonian mechanics; its change in the course of
> time is completely determined by the quantum mechanical equation, but it does not
> allow a description in space and time. The observation, on the other hand, enforces
> the description in space and time but breaks the determined continuity of the
> probability function by changing our knowledge of the system.
>
> *(Heisenberg, 1958d, p. 50)*

It is evident from this passage that Heisenberg interprets the complementarity of
space-time and causal descriptions quite differently from Bohr. For Heisenberg,
as for Bohr, all experience must be described in the usual concepts of space and
time, and this meant a description using the concepts of classical physics –
position, time, momentum and energy, albeit within the limits specified by the
uncertainty relations. The unobserved system, on the other hand, can be
described only by the Schrödinger equation in multi-dimensional configuration
space. The transition from a causal description in terms of the ψ-function to a
classical space-time description is characterised by the discontinuous change
that occurs in the act of measurement. At this point, we shift from one mode of
description to another. The two modes of description are then said to be
complementary to one another.

Perhaps the closest Bohr ever came to a general definition of the idea of
complementarity can be found in his introductory survey to a series of essays
published in 1929, where he wrote that the quantum postulate 'forces us to
adopt a new mode of description designated as *complementary* in the sense that
any given application of classical concepts precludes the simultaneous use of
other classical concepts, which in a different connection are equally necessary
for the elucidation of phenomena' (Bohr, 1987a, p. 10). Heisenberg's extension
of the notion of space-time and causal complementarity as we have seen
represents a quite different view. This is expressed in his 1955 article where
he writes: 'The characterization of a [quantum] system by its Hilbert vector [or
by the ψ-function in configuration space] is *complementary* to its description in
terms of classical concepts' (Heisenberg, 1955, p. 27, emphasis added). This
divergence between the two physicists is in fact symptomatic of a more fundamental difference in their philosophical approach to interpreting quantum
mechanics. As Heisenberg later recalled, Bohr's approach to the problem of
interpretation had from the beginning been quite different from his own:

> [Bohr's] insight into the structure of the theory was not a result of mathematical analysis
> of the basic assumptions, but rather of intense occupation with the actual phenomena,
> such that it was possible for him to sense the relationship intuitively, rather than

derive them formally … Bohr was primarily a philosopher, not a physicist … I noticed that mathematical clarity had in itself no virtue for Bohr. He feared that the formal mathematical structure would obscure the physical core of the problem, and in any case he was convinced that *a complete physical explanation should absolutely precede the mathematical formulation*.

(Heisenberg, 1967b, pp. 95, 98, emphasis added)

As a number of scholars have pointed out, one of the key issues that separated the two physicists was Heisenberg's emphasis on the need to understand the meaning of the quantum formalism, in direct contrast to Bohr's search for a deeper understanding of the mutually exclusive conditions for a complete and exhaustive description of the object in quantum mechanics. Bohr and Heisenberg never successfully resolved their differences (Beller, 1999, p. 143). This is evident in their respective views of wave–particle duality discussed in Chapter 4.

6.3 Weizsäcker's reconstruction

Heisenberg's re-interpretation of the complementarity of space-time and the causal description in quantum mechanics has sometimes mistakenly been attributed to Bohr. This is perhaps best illustrated through a paper Weizsäcker published on the subject in 1955 (Weizsäcker, 1955, pp. 521–29, 545–55). Presented on the occasion of Bohr's seventieth birthday, the paper offered a careful analysis of Bohr's concept of complementarity, which Weizsäcker had long felt to be his most decisive contribution to the philosophy of physics. In his paper, Weizsäcker distinguished between two forms of complementarity, both of which he attributed to Bohr. 'The complementarity between the space-time description and the claim of causality', wrote Weizsäcker, 'is precisely the complementarity between the description of nature in terms of classical notions and in terms of the ψ-function' (Weizsäcker, 1955, pp. 525–6). Weizsäcker labelled this idea 'circular complementarity', while he coined the term 'parallel complementarity' to refer to the mutually exclusive relationship that existed between pairs of classical concepts such as position and momentum or energy and time (in different experimental arrangements). Circular complementarity was, according to Weizsäcker, Bohr's most penetrating and original insight into quantum mechanics. Such a view, however, as we have already seen, is based on a misinterpretation of Bohr's Como paper – a misinterpretation that owes its origin to Heisenberg.

After an exchange of correspondence with Bohr in 1955–6 following the publication of his paper, Weizsäcker came to the realisation that he had misunderstood Bohr's basic position (AHQP, BSC, Bohr to Weizsäcker, 20 December 1955; Weizsäcker to Bohr, 17 January 1956; Bohr to Weizsäcker, 28 January 1956; Bohr to Weizsäcker, 5 March 1956). In a later edition of Weizsäcker's book *Zum Weltbild der Physik* in which his article appeared, Weizsäcker appended a correction at the end of the paper, in which he explained that contrary to the view expressed in the paper, the 'complementarity between space-time description and causality' actually represents the complementary relationship between space-time coordination and momentum–energy conservation laws. 'Since I have to admit that I misunderstood Bohr in these points it is questionable whether the expression circular complementarity should be maintained' (Weizsäcker, 1963, p. 330). In his interview with Kuhn in 1963, Weizsäcker explained his mistake in interpreting Bohr's meaning in referring to the 'complementarity between a description in space-time concepts and a description by causal concepts'. As Weizsäcker explained, it had initially appeared to him that according to Bohr,

> the description in space-time concepts was a description by classical physics while the other [causal description] was a description by the Schrödinger wave function which obeys a differential equation. *This I had taken from Heisenberg's 1931 book …* Bohr then wrote to me [in 1956] that this was wrong, that I misunderstood him here, because a description by space-time concepts to him meant just a description of the concepts by time and position, and the description in the frame of causality to him meant a description according to conservation laws.
> *(AHQP, Interview with Weizsäcker, 9 June 1963, p. 26, emphasis added)*

It is significant that Weizsäcker here cites Heisenberg's book *The Physical Principles of the Quantum Theory* as the source of his own misunderstanding of Bohr's concept of complementarity. Indeed, in his interview with Kuhn, Weizsäcker explained: 'The first mistake was Heisenberg's, but I think it was a very natural mistake' (AHQP, 9 June 1963, p. 26). In their analyses of the concept of complementarity, Michel Bitbol, Carsten Held and Max Jammer have made reference to Weizsäcker's misunderstanding of Bohr's complementarity, but nowhere is the original misinterpretation attributed to Heisenberg (Jammer, 1974, pp. 103–4; Held, 1994, p. 883; Bitbol, 1996, p. 219).

It remains here to comment on whether Heisenberg's reinterpretation of Bohr's notion of complementarity was based on a misunderstanding or was the result of deliberate obfuscation of Bohr's own views. This question is difficult to answer, as there is little direct evidence in favour of either view. Mara Beller has argued that physicists such as Bohr, Heisenberg and Born – from the 'orthodox' camp – often deliberately concealed their more philosophical differences with one another in an effort to present the impression of unity and agreement (Beller, 1999). There seems to be some indication that by the mid-1930s this was the case. In the case of complementarity, however, it appears more plausible that Heisenberg originally misinterpreted Bohr's formulation as a result of the somewhat ambiguous wording of the crucial passage in Bohr's Como lecture.

Scholars have long recognised that understanding precisely what Bohr meant in his published writings is an immensely difficult task (Stapp, 1972, p. 1098; Faye & Folse, 1994, pp. xxvi–xxvii). In this context, we should note that many of Bohr's contemporaries, even those who were sympathetic to his ideas, also found his writings elusive and urged him to express himself more clearly in print. This is true of Pauli, who suggested to Bohr on 13 January 1928, that the notion of 'the complementarity of causal and space-time description requires further elucidation' (Bohr, 1985, p. 41). In an attempt to more carefully articulate his views, in 1929 Bohr temporarily abandoned the term 'complementarity' and replaced it with the term 'reciprocity' (Bohr, 1987b, p. 95). However, soon after the paper was published, Bohr conceded to Pauli that 'the change of name was a blunder' and subsequently returned to his original term 'complementarity' (Bohr to Pauli, 29 July 1929, Bohr, 1985, p. 195). Given the general state of confusion that surrounded Bohr's notion of complementarity, it seems likely that Heisenberg's interpretation, at least initially, stems from a mistaken reading of the critical passage in question, rather than an attempt to conceal his disagreement with Bohr's own interpretation. However, Heisenberg's later writings suggest that by the 1950s he had became more aware of the extent to which his own views were in conflict with Bohr's.

6.4 Mutually exclusive experimental arrangements

Though Bohr's early illustrations of complementarity focused more on the concept of stationary states, he did not limit it to such an analysis. In discussions with Heisenberg in 1927, Bohr argued that the uncertainty relations for the free particle were also a special case of the more general viewpoint of complementarity which centred on the mutual exclusivity of experimental arrangements. Heisenberg did come to accept this view, though as will become clear, he and Bohr adopted quite different underlying philosophical attitudes to this form of kinematic-dynamic complementarity.

In the later sections of the 1927 Como paper, Bohr argued that 'the measurement of the positional coordinates of a particle is accompanied not only by a finite change in the dynamical variables, but also the fixation of its position means a complete rupture in the causal [momentum–energy] description of its dynamical behaviour'. Conversely, 'the determination of its momentum' by measurement means that we cannot pinpoint the position of the particle in space. Bohr concluded that it was just this situation that illustrated 'most strikingly the complementary character of atomic phenomena' (Bohr, 1928, p. 584). In a letter to Orseen in November 1928, Bohr explained:

> In order to be able to say that a particle at a given time has been at a given position, we must know that some more closely specified diaphragm has been opened and closed at specific times. It is now easy to demonstrate that the uncertainty in the description ... corresponds exactly to our ignorance regarding respectively the momentum exchange between the individual and the rigid bodies, to which the diaphragm is fixed, and the energy exchange with the machinery or the clock responsible for the opening and the closing. For the space-time determination, we evidently pay by the rupture in the momentum–energy description.
>
> *(Bohr, 1985, p. 190)*

In Pauli's classic article for the *Handbuch der Physik*, published in 1933, he explained that 'in order to determine the position and momentum of a particle' it is necessary to employ '*mutually exclusive experimental arrangements*'. On the one hand, the determination of the position of a particle requires that we use 'spatially fixed apparatuses (scales, clocks, screens)', which prevent us from knowing the precise magnitude of the momentum transferred to the apparatus. 'On the other hand, the determination of momentum would prevent the pinpointing of the particle in space and time' (Pauli, 1980, p. 7, emphasis in original). Reflecting on this point of view, Heisenberg declared that: 'The complementary relationship between different aspects of the same physical process is indeed characteristic for the whole structure of quantum mechanics' (Heisenberg, 1965, p. 299). Indeed, in a letter to Moritz Schlick in November 1932, Heisenberg was critical of the recent works of Franck and Reichenbach, who had 'hardly mentioned the real point of quantum theory, namely Bohr's complementarity', and instead had only chosen to 'reproduce the more superficial aspects in Born's papers and mine' (Heisenberg to Schlick, 21 November 1932, Mehra & Rechenberg, 2001, p. 692). As Heisenberg put it in his 1955–6 lectures, 'knowledge of the position of a particle is complementary to the knowledge of its velocity or momentum' (Heisenberg, 1958d, p. 50).

In his Nobel speech of 1933, Heisenberg argued that 'we are indebted to Bohr' for his 'concept of complementarity' which provides 'the clearest analysis of the conceptual principles of quantum mechanics' (Heisenberg, 1965, p. 299). Yet, Heisenberg and Bohr adopted fundamentally different philosophical attitudes in their understanding of complementarity. These differences are best illustrated by considering their respective responses to the challenge to quantum mechanics published in 1935 by Einstein, Podolsky and Rosen in which the authors argued that the ψ-function gives only an incomplete description of reality, and expressed the hope that a more complete description of the motion of a particle would, in time, be discovered. In the early phase of the Einstein–Bohr exchange, the debate centred around whether there are in fact definite limits on the accuracy of measurement. This argument was to become central in Bohr's ongoing debate with Einstein over the interpretation of quantum mechanics, perhaps most notably at the sixth Solvay conference in 1930. During this time, Bohr's work was devoted to thwarting Einstein's attempts to circumvent the uncertainty relations, by finding a means of determining the position and momentum of a particle simultaneously, famously exemplified in the photon-in-box thought-experiment at the Solvay conference in 1930. As Bohr later recalled, 'The problem raised by Einstein was now to what extent a control of the momentum and energy transfer, involved in the location of the particle in space and time, can be used in the further specification of the state of the particle' (Bohr, 1949, p. 215). However, after 1935, Einstein adopted a different approach. In the Einstein–Podolsky–Rosen paper (Einstein *et al.*, 1935), and in his later writings, Einstein took the view that quantum mechanics does not preclude the possibility that the '(free) particle really has a definite position and a definite momentum, even if they cannot both be ascertained by measurement in the same individual case' (Einstein, 1971, p. 169). In the EPR paper the authors describe a scenario where two spatially separated systems, which had previously interacted, cannot be assigned a well-defined kinematic-dynamic state according to quantum mechanics until an observation is made. Einstein, Podolsky and Rosen argued that one could nevertheless infer from measurements carried out on one of the systems, the state of the other spatially separate system, thereby proving that quantum mechanics in its current form was essentially an incomplete theory.

Bohr's carefully considered response to this challenge in his reply to the EPR paper was to point out that the mutually experimental arrangements serve to *define the very conditions* under which we can unambiguously employ such concepts as position and momentum. As many scholars have pointed out, this emphasis on the conditions for the *unambiguous* use of concepts lies at the heart of Bohr's response to Einstein's challenge to quantum mechanics in the 1930s and 1940s.[2] In a paper presented in 1936 at the 'Unity of Science' conference in Copenhagen, Bohr explained that the statement that 'the position and momentum of a particle cannot simultaneously be measured with arbitrary accuracy' is somewhat misleading. It is not simply that we cannot *measure* the two

---

[2] As Murdoch puts it, 'After 1935 Bohr expressed the indefinability thesis in what may be called "semantic" as distinct form "ontic" terms' (Murdoch, 1987, p. 145). In a similar vein Jan Faye writes: 'After 1935 his grounds for asserting complementarity were not so much epistemological as they were conceptual or semantical' (Faye, 1991, p. 186).

well-defined attributes of an object, but rather that 'the whole situation in atomic physics deprives of all meaning such inherent attributes as the idealizations of classical physics would ascribe to the object'. In order to determine the position and momentum of an electron, it is necessary to 'use two different experimental arrangements, of which only one permits the *unambiguous use* of the concept of position, while only the other permits the *application* of the concept of momentum defined as it is solely by the law of conservation' (Bohr, 1937, p. 293, emphasis added). The key point for Bohr is that the concepts of position and momentum only have a well-defined applicability in mutually exclusive experimental arrangements. Again in 1948, Bohr emphasised the fact that the different experimental arrangements provide the 'the mutually exclusive conditions for the *unambiguous use* of space-time coordination, on the one hand, and dynamical conservation laws, on the other' (Bohr, 1948, p. 315). Through the complementary description, we learn the previously 'unrecognized presuppositions for an *unambiguous use* of our most simple concepts' (Bohr, 1937, pp. 289–90). John Honner has argued that 'Bohr's principal concerns were of a kind which, since Kant, have been commonly described as transcendental', that is to say Bohr's work expresses 'a fundamental concern with the necessary conditions for the possibility of (experiential) knowledge' (Honner, 1982, p. 1). The emphasis on the experimental conditions for the unambiguous use of concepts lies at the heart of Bohr's response to Einstein's challenge to quantum mechanics.

Bohr's emphasis on the non-ambiguity of the complementary description has been brought to light in a number of recent works on Bohr's philosophy of quantum mechanics. Catherine Chevalley, for example, has emphasised that non-ambiguity 'is an essential notion in the epistemology of Bohr'. Only an 'unambiguous description' would serve as 'an objective description' of the phenomena (Chevalley, 1991a, p. 351). Yet, Heisenberg appears to have departed from Bohr's view. Heisenberg was rather more skeptical about the possibility of removing the ambiguity from a quantum-mechanical description using classical concepts. As he told Stapp in 1972, 'it may be a point in the Copenhagen interpretation that its language has a certain degree of vagueness, and I doubt whether it can become clearer by trying to avoid this vagueness' (Stapp, 1972, p. 1113). In a remarkable passage in his *Physics and Philosophy*, Heisenberg explicitly took issue with Bohr's claim that complementarity had brought about an unambiguous description of atomic phenomena:

> One may say that the concept of complementarity introduced by Bohr into the interpretation of quantum theory has encouraged the physicists to use an *ambiguous*, rather than unambiguous language, to use the classical concepts in a somewhat *vague*

manner in conformity with the principle of uncertainty, to apply alternatively different classical concepts [such as position and momentum] which would lead to contradictions if used simultaneously. In this way one speaks about electronic orbits, about matter waves and charge density, about energy and momentum, etc., always conscious of the fact that these concepts have only a very limited range of applicability. When this *vague* and *unsystematic* use of language leads to difficulties, the physicist has to withdraw into the mathematical scheme.

(Heisenberg, 1958d, pp. 154–5, emphasis added)

Heisenberg's view that Bohr's concept of complementarity has encouraged a 'vague' and 'ambiguous' use of language stands in sharp contrast to Bohr's repeated claims that complementarity enables us to uncover the previously 'unrecognized presuppositions for an *unambiguous use* of our most simple concepts' (Bohr, 1937, p. 290). Bohr had envisaged complementarity as a means of overcoming the fundamental ambiguity in the description of nature, whereas Heisenberg saw it as an illustration of the ambiguity inherent in the necessity of using classical concepts in quantum mechanics. It would therefore seem that Heisenberg's retreat into the formalism of quantum mechanics, and his redefinition of the notion of complementarity to include the causal description in terms of the ψ-function, expresses a fundamentally different attitude to that of Bohr regarding the place of complementarity in the interpretation of quantum mechanics.

Bohr's response to Einstein's charge of the 'incompleteness' of quantum mechanics is grounded in his analysis of the conditions under which concepts such as position and momentum can be meaningfully applied to an object in quantum mechanics (Bohr, 1935, 1937, 1948, 1949). This situation, Bohr maintained, is faithfully mirrored in the formal structure of quantum mechanics, according to which it is impossible to specify simultaneously the position and momentum of a particle. Thus, Bohr's defense of the 'completeness' of quantum mechanics did not directly engage with a formal analysis of the theory, but attempts to show that the epistemological situation we are confronted with is entirely consistent with the laws of quantum mechanics. By contrast to Bohr's approach, Heisenberg never seems to have invoked the need for a complementary description as the justification for the 'completeness' of quantum mechanics. As Melanie Frappier has argued, Heisenberg's defence of the 'completeness' of quantum mechanics ultimately rested on his notion of a 'closed theory', which he defined as 'an internally consistent set of axioms and definitions expressed in mathematical equations representing the relations existing between the theory's concepts' (Frappier, 2004, p. 164). In discussions at the *Science and Synthesis* Colloquium in 1965, Heisenberg emphasised that in his view the theory of non-relativistic quantum mechanics can be regarded as 'complete' in the sense that it exhibits all the characteristics of a 'closed theory' in physics. That is to say, in the region where we can and do employ *concepts* such as stationary states and transition probabilities, the *laws* of quantum mechanics are exact and cannot be improved upon (Maheu, 1971, pp. 144–5). For Heisenberg, the laws of quantum mechanics are *valid* and therefore in some sense *complete*, albeit within a limited domain of applicability. A deeper examination of Heisenberg's argument for the completeness of quantum mechanics requires that we look closely at his conceptualisation of the object–instrument divide or 'cut' [*Schnitt*], which he articulated in the 1930s. It is to this that we now turn our attention.

6.5 The object–instrument divide

Complementarity, in Bohr's view, was intimately connected with the experimental conditions under which a quantum object is observed. Early on, Bohr recognised that the classical concept of an 'object' which has a well-defined 'state' and which interacts with a measuring instrument is rendered problematic in quantum mechanics. This is grounded in the fact that, as Max Born observed in 1926, in quantum mechanics 'one cannot, as in classical mechanics, pick out a state of one system and determine how this is influenced by a state of the other system since all states of both systems are coupled in a complicated way' (Born, 1983, pp. 52–3). In other words, the common notion of an object as having an independent, separate dynamical state which is modified in the passage of time by its interaction with other things (e.g. field, particles, and instruments) can no longer be accepted uncritically in quantum mechanics. If we treat the interaction between an electron and measuring instrument quantum-mechanically, we must employ a wave function in multi-dimensional configuration space, in which case, we can no longer delineate the 'object' from the 'instrument'. They form part of a single 'quantum-mechanical system'. In 1935, Schrödinger coined the term 'entanglement' for this non-classical feature of quantum systems. This is precisely what had most troubled Einstein about quantum mechanics in the 1930s and 1940s (Einstein, 1971, pp. 168–73; Howard, 1985).

As Bohr noted, this paradoxical situation in quantum mechanics has serious implications for the concept of observation. In order to observe a quantum system, we must interact with it using a measuring instrument. This may be either an optical instrument such as the gamma-ray microscope, or some other device such as a photographic plate. Strictly speaking, such an interaction destroys the separability of the object and the instrument, since together they form a single quantum-mechanical system. For Bohr, this lay at the heart of the epistemological paradox of quantum mechanics. For, in order to observe an electron, for example, we must *assume* that the electron possesses an independent dynamic state (momentum, energy, etc.), which is in principle distinguishable from the state of the objects with which it interacts. This idea, which is embodied in the classical physical description, is for Bohr, the fundamental condition for the possibility of observation. Bohr regarded this condition as a simple logical demand, because, without such a presupposition, an electron cannot be an 'object' of empirical knowledge at all. As he explained at the 'Unity of Science' congress in 1936:

> [A] still further revision of the problem of observation has since been made necessary by the discovery of the universal quantum of action, which has taught us that the whole mode of description of classical physics, including the theory of relativity, retains its adequacy only as long as all quantities of action entering into the description are large compared to Planck's quantum ... This circumstance, at first sight paradoxical, finds its elucidation in the recognition that in this region [where classical mechanics breaks down] it is no longer possible sharply to distinguish between the autonomous behavior of a physical object and its inevitable interaction with other bodies serving as measuring instruments, the direct consideration of which is excluded by the very nature of the concept of observation itself.
>
> *(Bohr, 1937, p. 290)*

It is not merely that the act of measurement influences or disturbs the object of observation, but that it is no longer possible to distinguish between the 'object' and the 'measuring instrument' with which it interacts. A quantum-mechanical treatment of the observational interaction would paradoxically make the very distinction between object and instrument ambiguous. However, such a distinction is a necessary condition for empirical inquiry. As Bohr was to put it, only so long as we neglect the entanglement 'between the object and the measuring instrument, which unavoidably accompanies the establishment of any such connection' can we 'speak of an autonomous space-time behaviour of the object under observation' (Bohr, 1937, p. 291). To speak of an interaction between object and measuring instrument at all is to speak in terms of classical physics.

This argument made an immediate and lasting impression on Heisenberg. As he observed, it follows then that from an epistemological point of view 'a peculiar schism in our investigations of atomic processes is inevitable' (Heisenberg, 1952c, p. 15). In the discussions at the Como conference in September 1927, Heisenberg explained that in 'quantum mechanics, as Professor Bohr has displayed, observation plays a quite peculiar role'. In order to observe a quantum-mechanical object, 'one must therefore cut out a partial system somewhere from the world, and one must make "statements" or "observations" just about this partial system' (Bohr, 1985, p. 141). The existence of 'the cut between the observed system on the one hand and the observer and his apparatus on the other hand' is a necessary condition for the possibility of empirical knowledge. Without the assumption of such a divide we could not speak of the 'object' of empirical knowledge in quantum mechanics. Heisenberg emphasised the significance of the 'cut' [*Schnitt*] throughout the 1930s. In his lecture on 'Questions of Principle in Modern Physics' delivered in November 1935 in Vienna, Heisenberg explained:

> In this situation it follows automatically that, in a mathematical treatment of the process, a dividing line must be drawn between, on the one hand, the apparatus which we use as an aid in putting the question and thus, in a way, treat as part of ourselves, and on the other hand, the physical systems we wish to investigate. The latter we represent mathematically as a wave function. This function, according to quantum theory, consists of a differential equation which determines any future state from the present state of the function ... The dividing line between the system to be observed and the measuring apparatus is immediately defined by the nature of the problem but it obviously signifies no discontinuity of the physical process. For this reason there must, within certain limits, exist complete freedom in choosing the position of the dividing line.
>
> *(Heisenberg, 1952a, p. 49)*

This point had been emphasised in his lecture the previous year, in which Heisenberg argued that 'there arises the necessity to draw a clear dividing line in the description of atomic processes, between the measuring apparatus of the observer which is described in classical concepts, and the object under observation, whose behaviour is represented by a wave function' (Heisenberg, 1952c, p. 15). This was the central theme in an unpublished paper, written in 1935, entitled *Ist eine deterministische Ergänzung der Quantenmechanik möglich?*, in which Heisenberg outlined his own response to the criticisms of quantum mechanics which had emerged from such physicists as von Laue, Schrödinger and Einstein in the 1930s. A draft of the paper is contained in his letter to Pauli on 2 July 1935 (Pauli, 1985, pp. 409–18 [item 414]).[3] The paper, which was written at Pauli's urging, argued that a deterministic completion of quantum mechanics is in principle impossible. This is because in quantum mechanics we must draw a cut, between the quantum mechanical system to be investigated, represented by a wave function in configuration space, and the measuring instrument described by means of classical concepts. The critical point, for Heisenberg is that 'this cut can be shifted arbitrarily far in the direction of the observer in the region that can otherwise be described according to the laws of classical physics', but of course, 'the cut cannot be shifted arbitrarily in the direction of the atomic system' (Heisenberg, 1985b, p. 414). No matter where we chose to place the 'cut', classical physics remains valid on the side of the measuring device, and quantum

---

[3] Elise Crull, a PhD student in the History and Philosophy of Science program at the University of Notre Dame, has recently completed an English translation of Heisenberg's 1935 EPR paper.

mechanics remains valid on the side of the atomic system. Any attempt to supplement the existing theory of quantum mechanics with new parameters to give a 'more complete' theory would mean that the cut could no longer be shifted in the direction of the observer, thereby destroying this crucial feature of the quantum-mechanical description.

Heisenberg's defence of the completeness of quantum mechanics turns out to be rather different to the one expressed by Bohr in his response to the EPR paper. Writing to Pauli in July 1935, Heisenberg drew attention to the fact that Bohr's reply to the Einstein–Podolsky–Rosen challenge was 'very different from the thoughts that one might express in connection with the "Cut"', and it was for this reason that he expressed his desire to 'deal with the question of the incompleteness of quantum mechanics' from the perspective of the 'cut' between quantum object and measuring apparatus (Heisenberg to Pauli, 2 July 1935, Pauli, 1985, p. 408 [item 414]). The crucial point for Heisenberg was that the location of the cut 'cannot be established physically, and moreover it is precisely the arbitrariness in the choice of the location of the cut that is decisive for the application of quantum mechanics' (Heisenberg, 1985b, p. 416). As he explained to Bohr, without such a presupposition one would have to conclude that there exist 'two categories of physical systems – classical and quantum-mechanical ones' (AHQP, Heisenberg to Bohr, 29 September 1935). The classical/quantum divide is therefore not to be understood *ontologically* – as implying that there exist two kinds of entities in the world – but rather it is *epistemological* in nature – such a dividing line is a condition for the possibility of empirical knowledge (AHQP, BSC, Heisenberg to Bohr, 29 September 1935). Heisenberg acknowledged that we may of course chose to treat the object and the instrument as a single quantum-mechanical system to be represented as a wave function in configuration space, but that this composite system is then no longer accessible to us as an object of empirical knowledge. Indeed, as he pointed out in the discussions that followed Bohr's Como paper in September 1927: 'One may treat the whole world as *one* mechanical system, but then only a mathematical problem remains while access to observation is closed off' (Bohr, 1985, p. 141). The ψ-function, as Heisenberg put it in 1955, does 'not refer to real space or to a real property; it thus, so to speak, contains no physics at all' (Heisenberg, 1955, pp. 26–7).

Heisenberg appears to have followed Bohr closely in insisting that the measuring device must be described classically if it is to serve as a measuring instrument. As Weizsäcker was to put it, for Heisenberg, 'the classical description of instrument just meant that only so far as it was an approximation would the instrument be of use *as* an instrument' (Weizsäcker, 1987, p. 283). There were, however, key differences, which have often been overlooked, between Bohr and Heisenberg in their respective analyses of the quantum–classical divide. We know from an exchange of correspondence in autumn 1935 that Bohr objected to Heisenberg's view that the cut could be shifted arbitrarily far in the direction of the apparatus (AHQP, BSC, Heisenberg to Bohr, 10 August 1935, Bohr to Heisenberg, 10 September 1935). In Heisenberg's account the division between the quantum and the classical realms seems to coincide exactly with the object–instrument and microscopic–macroscopic distinction. This treatment of the cut has often been taken as a faithful representation of Bohr's most carefully considered view of the problem of measurement. However, Heisenberg frequently alluded to his disagreement with Bohr on this matter in the 1930s. In *Physics and Philosophy*, he explained that 'Bohr has emphasized that it is more realistic to state that the division into the object and rest of the world is not arbitrary' but is determined by the very nature of the experiment (Heisenberg, 1958d, p. 56). Writing to Heelan in 1975, Heisenberg explained that he and Bohr had never really resolved their disagreement about 'whether the cut between that part of the experiment which should be described in classical terms and the other quantum-theoretical part had a well defined position or not'. Here Heisenberg restated the point that for him the position of the 'cut could be moved around to some extent while Bohr preferred to think that the position is uniquely defined in every experiment' (Heelan, 1975, p. 137). Weizsäcker also later recalled that Heisenberg had disagreed with Bohr over the cut in the 1930s (Weizsäcker, 1987, p. 283). After their exchange of correspondence in August–September of 1935, Heisenberg eventually decided not to submit his paper for publication, and remained content to defer to Bohr's reply to EPR. Yet the difference of opinion was never fully resolved (Bohr to Heisenberg, 15 September 1935, Heisenberg to Bohr, 29 September 1935). In fact Heisenberg repeated the same argument, albeit in a somewhat abridged form, in his Vienna lecture of November later that year (Heisenberg, 1936).

The Heisenberg–Bohr exchange would seem to suggest that for Bohr, the quantum–classical distinction corresponds to something 'objective', and is not merely an arbitrary division. Howard has argued that a careful reconstruction of Bohr's views shows that for him 'the classical/quantum distinction' did not exactly coincide with the 'instrument/object distinction' (Howard, 1994b, pp. 203–11). Bohr, it seems, was keen to avoid the mistaken impression that a 'classical' instrument somehow interacts with the 'quantum-mechanical' object. According to Howard, 'Bohr required a classical description of some, but not necessarily all, features of the instrument and more surprisingly, per-haps, a classical description of some features of the observed object as well' (Howard, 1994b, p. 203). Thus, for Bohr, the separability between object and instrument pertains only to 'those properties of the measuring instrument [e.g. position] that are correlated, in the measurement interaction, with the properties of the observed object that we seek to measure'. Howard's reconstruction provides a new reading of many crucial passages from Bohr's writings, in which 'the classical/quantum distinction corresponds to an objective feature of the world' in the context of a particular experiment being carried out (Howard, 1994b, p. 211). While I do not have the space here to pursue a deeper analysis of this account of Bohr's doctrine of classical–quantum division, it seems to me that an appreciation of this point may shed some light on the reasons for Bohr's objections to Heisenberg's 'cut' argument in the mid-1930s.

The last three decades have seen the emergence of a growing body of literature on the history of quantum theory, which has undermined the commonly held view that the Copenhagen interpretation represents a unified point of view. Indeed Howard has gone so far as to suggest that 'until Heisenberg coined the term in 1955, there was no unitary Copenhagen interpretation of quantum mechanics'. All we had was 'a group of thinkers' determined 'to defend quantum mechanics as a complete and correct theory' (Howard, 2004, p. 680). Although Heisenberg's writings in the 1930s give the impression that he shared Bohr's general viewpoint of complementarity, Heisenberg's interpretation differed in crucial respects from Bohr's. Heisenberg interpreted Bohr's assertion that space-time and causal descriptions were complementary as meaning that a description in terms of the deterministic evolution of the wave function was mutually exclusive to a space-time description of the observed phenomena, though both were necessary. This was quite different from the interpretation that had been presented by Bohr, according to which the conditions for determining the energy or momentum of a system were mutually exclusive to those required for a representation in space and time.

As I have argued, the critical divergence between Heisenberg and Bohr over the meaning of complementarity is due to Heisenberg's misreading of the crucial passage in Bohr's famous Como paper. One the one hand, Heisenberg felt it was necessary to describe an isolated quantum-mechanical system in terms of the Schrödinger equation, but, on the other hand, as Bohr had so often pointed out, the results of all experiments must ultimately be expressed in terms of classical concepts. In this sense, Heisenberg argued that these two modes of description, which he regarded as mutually exclusive but complementary, were essential for the physical and epistemological interpretation of quantum mechanics.

Secondly, while accepting Bohr's view that the mutually exclusive experimental arrangements were indispensable for measuring the electron's position and momentum, Heisenberg never appears to have followed Bohr's interpretation of kinematic-dynamic complementarity, according to which such experimental arrangements serve to define the very conditions for the *unambiguous* use of classical concepts. By contrast, Heisenberg stressed the inevitable *ambiguity* inherent in the use of such concepts in quantum mechanics. To this extent, Heisenberg never seems to have understood the sense in which Bohr held the general 'viewpoint' or the 'framework' of complementarity to be *necessary* and *fundamental* in quantum mechanics. He did, however, remain wedded to Bohr's view that classical concepts, while having only limited applicability, are indispensable in the description of experience. This view is in fact the central plank of Heisenberg's later epistemology. As we shall see in the next chapter, it is a distinctly neo-Kantian concern with the *conditions for the possibility of empirical knowledge* that turns out to lie at the heart of Heisenberg's interpretation. Only once we have understood this aspect of Heisenberg's thought can we then understand his later efforts in the 1950s to develop a new concept of reality in quantum mechanics.

PART III

Heisenberg's epistemology and ontology
of quantum mechanics

7

The transformation of Kantian philosophy

Much has been written on the underlying Kantian impulses in Bohr's philosophy of complementarity (Weizsäcker, 1971a; Hooker, 1972; Folse, 1978; Honner, 1982; Kaiser, 1992). Yet, scholars have largely ignored Heisenberg's philosophy of quantum mechanics and its relationship to Kant, despite the fact that Heisenberg devoted considerable attention to precisely this question. Although Heisenberg entertained a moderate interest in epistemological problems throughout the 1920s, it was not until the 1930s upon the completion of quantum mechanics that he turned his attention to probing in more depth the epistemological lesson of quantum mechanics. Here Kant loomed large. This chapter examines the way in which Heisenberg appropriated and modified the Kantian philosophy into his own interpretation of quantum mechanics. More specifically, I argue that after 1930, Heisenberg interpreted Bohr's doctrine of classical concepts through his own transformation of Kant's concept of the *a priori*. Here, space, time and causality remain, for Heisenberg, the conditions for the possibility of experience, but unlike Kant, for Heisenberg, they do not have transcendental universality and necessity. Rather, such concepts have arisen through the historical development of human language, and turn out to have only limited range of applicability. To this extent, Heisenberg defined such concepts as 'practically *a priori*'. This was a view to which Heisenberg would remain committed.

Here, one can see two decisive influences on Heisenberg's thought: first, Bohr's analysis of observation as essentially a classical concept, and secondly, the epistemological discussions in Leipzig with Weizsäcker and Hermann on the significance of Kant's philosophy for quantum mechanics. As discussed in the previous chapter, Bohr had argued that although quantum mechanics precludes us from drawing a sharp demarcation between the object of empirical knowledge and the measuring instrument, the very concept of observation demands that we must draw such a dividing line. Furthermore, all experience must be described in terms of the concepts of *classical physics*. Meanwhile, the 1920s witnessed a series of attempts by philosophers such as Reichenbach, Schlick, Carnap, Cassirer, Weyl and Husserl, to modify and refine Kant's distinction between a priori and a posteriori judgements, albeit in different ways. Logical positivism, neo-Kantianism and phenomenology all offered different solutions to the Kantian problem of knowledge. Bringing Bohr's analysis into line with this broader neo-Kantian philosophical project of clarifying the meaning of the a priori, Heisenberg arrived at a distinctive view of the conditions for the possibility of knowledge of the atomic world. Heisenberg sometimes referred to his position as involving a 'practical' transformation of Kantian philosophy.[1]

While my focus in this chapter is primarily on Heisenberg, in order to bring into sharper focus certain aspects of Heisenberg's thought, I have drawn explicitly on the published writings of Bohr and Weizsäcker, both of whom were in close contact with Heisenberg during the late 1920s and 1930s. Yet, as will become evident, Heisenberg adopted a unique and original perspective on the neo-Kantian interpretation of quantum mechanics, which is worthy of examination not only from a historical, but also from a philosophical, point of view. Heisenberg's reflections on the conditions for the possibility of knowledge in modern physics offer a new insight into one of the enduring themes of twentieth century philosophy, which for lack of a better term, we might refer to as the 'relativisation' of the a priori. Furthermore, in clarifying the neo-Kantian dimension of Heisenberg's philosophy of quantum mechanics, we can see more clearly the way in which Heisenberg's later effort to develop a quantum ontology through his introduction of the concept of *potentia* was grounded in his transformation of Kantian epistemology.

7.1 Heisenberg's early confrontation with Kantian philosophy

Heisenberg's philosophical writings reveal a central preoccupation with Kant. In the late 1920s, Heisenberg turned his attention to the relationship between critical philosophy and the interpretation of quantum mechanics, to which he himself had contributed decisively. Heisenberg's student and close friend, Carl Friedrich von Weizsäcker, recalls that it was evident from their

---

[1] Heisenberg's views should not be confused with the kind of 'pragmatic Kantianism' espoused by such thinkers as C.I. Lewis, Quine and the later Wittgenstein, though there are some similarities (Lewis, 1929).

discussions in Leipzig during the 1920s that he 'had spent some time studying Kant' (AHQP, Interview with Weizsäcker, 9 June 1963, p. 2). Though Heisenberg certainly admired Kant's penetrating philosophical insight, he was not a Kantian in any strict sense of the term, and took the attitude that philosophy must move beyond critical idealism.

The first reference to Kantian philosophy in Heisenberg's work appears in his 1928 lecture on 'The Epistemological Problems of Modern Physics' presented shortly after his professorial appointment in Leipzig (Heisenberg, 1984a). There, he presented a critique of Kantian philosophy from the perspective of the recent developments in modern physics. Heisenberg declared that the theories of relativity and quantum mechanics had fundamentally transformed physics in the last three decades and in doing so had rendered problematic Kant's distinction between a priori and a posteriori knowledge. Whereas for Kant, the axioms of Euclidean geometry are synthetic a priori judgements, and to this extent necessary and universal conditions for the possibility of experience, Einstein's general theory of relativity had demonstrated that space is not Euclidean. So too, Heisenberg felt that quantum mechanics had now dealt a serious blow to Kant's view that causality is an a priori condition for the possibility of knowledge. This was the focus of Heisenberg's lecture, 'The Causal Law and Quantum Mechanics' delivered in Königsberg in 1930, in which he argued that quantum mechanics forces us to abandon the principle of strict causality characteristic of classical physics (Heisenberg, 1931a).

In the period between 1928 and 1931, Heisenberg called for a renewed effort on the part of philosophers to 'unravel again the basic problem of Kant's epistemology' by drawing on the latest discoveries of modern physics (Heisenberg, 1984a, p. 28). Quantum mechanics, Heisenberg argued, no longer permits us to describe the object independent of its interaction with the measuring apparatus as existing in space and time and to this extent it has demolished the idea of 'a clear separation of subject and object' underpinning Kantian epistemology (Heisenberg, 1931a, p. 181). As Heisenberg spelled out: 'Whereas before hand the spatio-temporal description could be applied to an isolated object, it is now essentially linked to the interaction between the object and the observer or its apparatus. The object in total isolation no longer has, in essence, any describable properties' (Heisenberg, 1931a, p. 182). We must be careful here not to interpret Heisenberg as advocating some form of 'subjectivism'. When Heisenberg refers to the impossibility of a sharp separation between the subject and object in quantum mechanics, he does *not* mean 'the observer appears as a necessary part of the whole structure in his full capacity as a conscious being'. This view was taken by some physicists, notably Heitler and Wigner (Heitler, 1949, pp. 194–5; Wigner, 1983a; 1983b). Yet, this was not what Heisenberg had intended. Heisenberg's reference to the subject-relation in the context of quantum mechanics should be understood as pertaining to the *physical interaction* between the measuring instrument and the quantum object.

If we look carefully at Heisenberg writings in the early 1930s, we find that his early critique of Kantian philosophy centred around the very issues which were, at that time, confronting philosophers schooled in the neo-Kantian tradition. By challenging the necessity and universality of the concepts of Euclidean space, time and causality in physics, the discoveries of modern physics appeared to undermine Kant's view that all empirical knowledge was grounded in synthetic a priori judgements. Heisenberg's critique of Kant in this early period in many respects echoes the criticisms of thinkers like Reichenbach, Carnap and Schlick, who were instrumental in the rise of logical positivism in the late 1920s and 1930s (Schlick, 1979a). Yet, like empiricists such as Reichenbach and Schlick, Heisenberg did not altogether deny that there is some non-empirical dimension of human knowledge. While on the one hand he asserted that our knowledge of the world is in some sense 'substantially reducible to experience', on the other hand he suggested that the concepts of physics 'sharpen' or 'render more precise' [*verschärfen*] what is given in experience (Heisenberg, 1984a, pp. 27–8). In the conclusion to his 1928 lecture, Heisenberg left it an open question, as to how we are to reinterpret the Kantian distinction between a priori and a posteriori knowledge in the aftermath of the theory of relativity and quantum mechanics. This was the central epistemological problem, which Heisenberg now saw as confronting those concerned with modern physics:

> In earlier times it was assumed that one could neatly distinguish between the *content* of our thoughts and experiences, and the *form* of our thought – that very "*a priori* form [*Gestalt*]". This no longer appears to be correct, because it is evident that also the *form* of our thinking is intimately linked to everyday experiences; it still remains an unresolved issue in what way this takes place. It would be an immensely interesting, but also a very difficult task to unravel again the basic problem of Kant's epistemology, so to speak starting anew and to try again the separation of the question of how much of our knowledge is derived from experience, and how much from our capacity to think. Kant's demarcation is untenable, but is it possible to draw a new boundary? But this is your task, not that of the natural scientists who are only supposed to supply the material to enable you to continue the work.
> *(Heisenberg, 1984a, p. 28, emphasis in original)*

Heisenberg here articulates what he took to be the central problematic of post-Kantian epistemology: how we are to distinguish between the *content* and *form* of knowledge? As Heelan rightly points out, Heisenberg 'situates himself within the perspective of critical philosophy, but of a critical philosophy in crisis' (Heelan, 1965, p. ix). It is worth noting that the fundamental question here posed by

Heisenberg concerning the Kantian distinction between a priori and a posteriori knowledge is more or less the same question posed by a number of prominent thinkers during the 1920s in the German-speaking world. Indeed the *Erkenntnisproblem* (or problem of knowledge) that Heisenberg posed in 1927, should be viewed in the wider context of the philosophical schools of thought that arose in the 1920s and 1930s in response to the crisis of classical physics brought about by the theory of relativity. Logical positivism, Husserlian phenomenology and neo-Kantianism also constituted different philosophical approaches in attempting to navigate a path between the excesses of Kant's idealism and Mach's positivism.

In spite of the weakening position of critical idealism in the 1920s, it should be noted that Kant remained extremely influential in the German-speaking world. As scholars such as Albert Coffa, Michael Friedman, and Don Howard have all argued, Kantian epistemology exerted a considerable influence in the rise of logical empiricism (Coffa, 1991; Friedman, 1994; Howard, 1994a; Friedman, 1996). This was particularly evident in the writings of Hans Reichenbach, Moritz Schlick and Rudolf Carnap, all of whom wrote important treatises on the philosophy of space and time during this period (Schlick, 1917; Reichenbach, 1920; Carnap, 1922). As Howard explains, by the mid-1920s, Schlick, Carnap and Reichenbach had 'conceded the need for the constitutive element in human intellection' and to this extent they did not hold a positivist conception of geometry (Howard, 1994a, p. 47). In 1921, Moritz Schlick explained that 'the forming of concepts of physical objects' cannot be the result of 'mere sensations and perceptions', but 'unquestionably presupposes certain principles of ordering and interpretation' (Schlick, 1979b, pp. 323–4). However, Schlick rejected Kant's view that the forms of thought were a priori in the sense of having 'the property of adopeicticity (of universal, necessary and inevitable validity)'. The empiricist will not deny that there are constitutive principles in human knowledge, 'he will deny only that they are synthetic and *a priori* in the sense defined' by Kant (Schlick, 1979b, pp. 323–4). Reichenbach was inclined to retain the term *a priori* to refer to such constitutive principles. In his essay on relativity in 1922, he explained 'some constitutive principles must always be presupposed for the establishment of empirical knowledge', though such principles are to be regarded as either hypotheses or conventions. Thus, for Reichenbach, 'the *a priori* loses its apodictic character, but it retains the more important property of being "constitutive of objects"' (Reichenbach, 1996, p. 284).

We find a similar approach in the work of Carnap. Friedman has argued that Carnap's major work of this period *Die logische Aufbau der Welt*, 'although by no means entirely independent of the empiricist-positivist tradition, originates within a primarily Kantian and neo-Kantian philosophical context'. For Carnap, just as for Reichenbach and Schlick: 'Experience is to be constituted from sensation on the basis of forms imposed by thought, but these forms are increasingly deprived of the fixed, synthetic *a priori* character ascribed to them by Kant' (Friedman, 1996, p. 391). To this extent, Friedman argues that the rise of logical empiricism was associated with 'a profound transformation of the Kantian concept of synthetic *a priori* principles' (Friedman, 1999, p. xv).

The logical empiricists were not the only school of thought who sought to reconstruct Kantian epistemology. Throughout the 1920s, a number of strands of thought emerged in the German-speaking world, each of which adopted a different attitude to the reframing of the a priori–a posteriori distinction. In 1927, Hermann Weyl explained that the critical reaction to Kant had taken different directions in philosophy: 'The general philosophical development ... has since taken a course that led to a split of Kant's judgments *a priori* in two directions'. In one direction, we find a line of thought culminating 'in Husserl's phenomenology, in which the *a priori* is much richer than in the Kantian system', while in the other direction, a priori judgements are relegated to the status of constitutive principles of theory construction, 'which according to the most extreme point of view (Poincaré) rest on pure convention' (Weyl, 1949, p. 134). Aligning himself with the viewpoint of phenomenology, Weyl argued that Einstein's 'general theory of relativity does not altogether deny that there is something aprioristic to the structure of the extensive medium of the external world [space], but the line between the *a priori* and the *a posteriori* is drawn at a different place' (Weyl, 1949, p. 134). One finds a similar view in the work of Ernst Cassirer, who argued that while Einstein's theory of relativity had demolished the idea that Euclidean geometry is a priori, it had in fact reinforced the perspective of Kant's critical idealism insofar as space and time must be understood, not as objective realities, but rather as forms of intuition, and as such they remain the a priori conditions for the possibility of experience (Cassirer, 1923, pp. 412–13, 438–44).

The diverging departures from Kantian philosophy evident in the work of the logical empiricism of Carnap, Schlick and Reichenbach, the transcendental phenomenology of Husserl, and the neo-Kantianism of Cassirer are not presented here to indicate Heisenberg's preference for one school of thought or another. Rather, they form the philosophical milieu in which we must situate Heisenberg's foray into epistemology. With the completion of quantum mechanics in the late 1920s, Heisenberg saw the task that again confronted the philosopher as being to 'unravel again the basic problem of Kant's epistemology'. This was an echo of Sommerfeld's call for 'a new Kant' (Sommerfeld, 1927, p. 235). While Heisenberg saw the elaboration of a new epistemology as chiefly the business of philosophers, and not physicists, Bohr's analysis of the epistemological problem of quantum mechanics – in particular his analysis of the concept of observation in the late 1920s – provided the impetus for his own 'pragmatic' transformation of Kantian philosophy. It is therefore necessary to turn our attention to Bohr's analysis of the concept of the empirical object before examining Heisenberg's views on the historical and linguistic dimensions of a priori knowledge.

7.2 The doctrine of classical concepts

By 1927, Bohr had turned his attention to the philosophical foundations of quantum mechanics, which for him, were centred on an analysis of the concept of observation. The central idea was that observation itself is essentially a *classical concept*, a view which we discussed in the previous chapter. Bohr's view rested on two closely related lines of argument, both of which can be traced back to his 1927 Como paper. The first of these is that we cannot delineate the 'object' under investigation from the 'measuring instrument' if object–instrument interaction is described quantum-mechanically. They form a single quantum-mechanical system. Yet, as Bohr pointed out, we must distinguish between the 'object' under observation and the device serving as a measuring instrument in order to speak of an 'object of knowledge' at all. To this extent we must employ the concepts of classical physics which embody the idea of separability. Bohr drew attention to this point in his 1927 Como paper, where he argued that 'the distinction between object and agency of measurement', which is so characteristic of classical physics, is 'inherent in the very idea of observation' (Bohr, 1928, p. 584). For Bohr, this lay at the heart of the epistemological paradox of quantum mechanics. In order to observe an electron, for example, we must *assume*, contrary to quantum mechanics, that the electron possesses an independent dynamic state, which is in principle distinguishable from the state of the objects with which it interacts. This idea, which is embodied in the classical mode of description, is, for Bohr, the fundamental condition for the possibility of observation.

As we have already seen, this line of argument was taken up by Heisenberg. In a lecture on 'Questions of Principle in Modern Physics' in November 1935, Heisenberg emphasised that in quantum mechanics 'a dividing line must be drawn between, on the one hand, the apparatus which we use as an aid in putting the question and thus, in a way, treat as part of ourselves, and on the other hand, the physical systems we wish to investigate' (Heisenberg, 1952a, p. 49). As was discussed in the last section of the previous chapter, Heisenberg was keen to point out in his correspondence with Bohr during this time that we are free to chose, within certain limits, the position of the dividing line. Heisenberg acknowledged that we may of course choose to treat the object and the instrument as a single quantum-mechanical system to be represented as a ψ-function in configuration space, but we are then no longer left with an 'object' of empirical inquiry. In *Physics and Philosophy*, Heisenberg explained that 'classical physics is just that idealization in which we can speak about parts of the world [i.e. objects] without any references to ourselves. Its success has led to the general ideal of the objective description of the world'. In posing the question to what extent 'does the Copenhagen interpretation of quantum theory comply with this ideal', Heisenberg answered: 'One may perhaps say that quantum theory corresponds to this ideal as far as possible'. Quantum physics 'starts from the division of the world into the "object" and the rest of the world' but with the knowledge that the division is an arbitrary one (Heisenberg, 1958d, pp. 54–5). Heisenberg here recognised Bohr's point that we must draw a distinction between object and the agency of observation, even though such a distinction is, according to quantum theory, only an idealisation. It is only in this classical limit that we can speak of an independently existing object at all.

The second line of argument employed by Bohr, and subsequently taken up by Heisenberg, was that the actual results of observation must be described through the concepts of classical physics. This is because the interaction between an electron and a measuring instrument can only be described if we assume that at the point of interaction, the electron is 'in' space and 'has' momentum. Without this assumption, it is impossible to make sense of what has been registered in our experimental apparatus. As Bohr was to put it, 'it lies in the nature of physical observation ... that all experience must ultimately be expressed in terms of classical concepts' (Bohr, 1987b, p. 94). Bohr was categorical on this point, insisting that 'no matter how far the phenomena transcends the scope of classical physical description, all evidence must be given in terms of the concepts of classical physics' (Bohr, 1987b, p. 94). If we consider the appearance of a 'mark' on a photographic plate, this can only serve as an 'observation' of a particle if we can unambiguously infer something about the particle from it. In such a case, we must assume that a particle, say an electron, has impinged on the photographic plate at a specific area, that is to say the electron has 'interacted' with the measuring instrument through the exchange of momentum and energy somewhere in space and time. In this sense, we conceptualise the interaction in terms of classical physics.

Heisenberg was also in agreement with Bohr regarding the need to describe the results of measurement in terms of classical concepts. To quote Heisenberg: 'The behaviour of the observer as well as his measuring apparatus must therefore be discussed according to the laws of classical physics'. This is 'a necessary condition' which must hold 'before one can from the result of measurements, unequivocally conclude what has happened' (Heisenberg, 1965, p. 298). To this extent we must employ classical theory in accounting for the interaction between object and instrument. In elaborating on this point in his unpublished paper of 1935, Heisenberg argued that when we observe the blackening of a photographic plate, we must interpret this as having been caused by a causal chain of events leading back to the interaction of the particle and the plate:

> For our description of nature is based in the last analysis on the assumption that it is possible to speak of objective events in space and time. It would simply be impossible to speak of the correctness of the predictions of any theory if we did not start from the assumption that the appearance of a definite event is an objective fact that does not depend on our observation of this event.
>
> *(Heisenberg, 1985b, pp. 414–15)*

Furthermore, because the measuring instrument must be described in terms of classical physics, one must assume that the transfer of energy and momentum between the object and the instrument is governed by classical conservation laws. Without such an assumption, it would not be possible to infer the electron's momentum or energy from the measurement. To this extent, we must presuppose that the causal interaction between the 'object' and the 'instrument' can be described in the classical approximation (as a transfer of momentum and energy) in order to have *empirical knowledge* of the electron at all. As Heisenberg was to put it in later years, although a measuring instrument, like any other object, can be described by means of the equations of quantum mechanics, it serves as a measuring instrument 'only when the observations it yields enable us to arrive at unequivocal conclusions about the phenomena under observation, only when a strict causal connection [between object and instrument] can be assumed to exist' (Heisenberg, 1971, pp. 129–30).

Thus far Heisenberg's account of the doctrine of classical concepts appears to be more or less the same view we find in Bohr's work. However, there were in fact important differences in their respective philosophical viewpoints. Scholars such as Folse and Kaiser have pointed out that Bohr was in fact quite critical of Kant's thesis of a priori knowledge in formulating his own philosophy of complementarity (Folse, 1978; Kaiser, 1992, pp. 220–6). In a similar vein, Shimony writes that while Bohr's rejection of metaphysics was in certain respects 'reminiscent of the epistemological system of Kant', he did not share Kant's view 'concerning the structure of human knowledge, like the possibility of synthetic *a priori* judgements' (Shimony, 1983, p. 210). Heisenberg, by contrast, would place far more emphasis on the a priori dimension of empirical knowledge in his epistemology of quantum mechanics, and in so doing he reinterpreted the doctrine of classical concepts within, what might be broadly termed, a neo-Kantian framework. It is to this that we now turn our attention.

7.3 The constitutive dimension of language and the forms of intuition

David Cassidy has suggested that 'Heisenberg's more philosophical views on the Copenhagen doctrine evolved primarily out of conversations with his Leipzig student and close personal friend, Carl Friedrich von Weizsäcker, and with Grete Hermann' (Cassidy, 1992, p. 257). Hermann had been a student of the neo-Kantian philosopher Leonard Nelson in Göttingen, before completing her doctoral dissertation in Leipzig in 1935 on the philosophical foundations of quantum mechanics. The discussions between Heisenberg, Weizsäcker and Hermann in the early 1930s centred on the relevance of Kantian philosophy for quantum mechanics (Heisenberg, 1971, pp. 117–24; AHQP, Interview with Weizsäcker, 9 June 1963, p. 18). Hermann's work in the 1930s constitutes perhaps the most elaborate defence of the Kantian philosophical interpretation of quantum mechanics against the attacks of the logical positivists (Hermann, 1935a,b, 1937; Hermann, May & Vogel, 1937).[2] One of Hermann's central arguments was that despite the fact that one can only make statistical predictions in quantum mechanics, causality is preserved insofar as it was always possible to assign a definite cause to any event that had occurred in the past. To this extent, she argued that in a modified sense the Kantian category of causality remained a condition for the possibility of knowledge. Writing to Bohr in June 1934, Heisenberg mentioned that he had profited from extensive discussions with Weizsäcker and Hermann on 'the "general questions", about which we have learnt so much from you' (AHQP, Heisenberg to Bohr, 17 June 1934).

Of particular importance to Heisenberg and Weizsäcker in their discussions in Leipzig was Kant's concept of the *a priori*. In his lecture on the 'Recent Changes in the Foundations of Exact Science', delivered in 1934, Heisenberg explicitly referred to the way in which quantum mechanics had brought a new insight to the question of the a priori status of space, time and causality. There, he announced that 'modern physics has more accurately defined the limits of the idea of the *a priori* in the exact sciences, than was possible in the time of Kant'

---

[2] In 2001, Lena Soler presented a paper entitled 'The Convergence of Transcendental Philosophy and Quantum Physics: Grete Henry–Hermann's 1935 Pioneering Proposal' in Bremen. The paper was presented at the invitation of the Philosophische-Politische Akademie on the occasion of the Grete Henry–Hermann Celebration.

(Heisenberg, 1952c, p. 21). Here, Heisenberg framed Bohr's doctrine of classical concepts in a distinctly neo-Kantian context:

> Here the question raised by Kant, and much discussed ever since, concerning the *a priority* of the forms of intuition and categories ... has been put into new light. On the one hand it has been shown that our space-time form of intuition and the laws of causality are not independent of all experience, in the sense that they must remain for the rest of time essential constituents of every physical theory. On the other hand, as Bohr particularly has stressed, the applicability of these forms of intuition, and of the law of causality is the premise of every scientific experience even in modern physics. For we can only communicate the course and result of a measurement by describing the necessary manual actions and instrument readings as objective, as events taking place in the space and time of our intuition. Neither could we infer the properties of the observed object if the law of causality did not guarantee an unambiguous connection between the instrument and the object.
>
> *(Heisenberg, 1934, pp. 700–1)*[3]

On occasions, Bohr had expressed himself in such a way as to suggest a direct relationship to Kant. In his 'Introductory Survey' in 1929, for example, Bohr referred explicitly to space and time as 'forms of perception' or 'forms of intuition' (*Anschauungsformen*), which suggests a link to Kant (Bohr, 1987a, p. 5). However, this was a connection that Bohr himself never explicitly acknowledged. Indeed, Bohr was more inclined to emphasise the point that his own view differed sharply from Kant's (Folse, 1985, pp. 216–19). Yet, both Weizsäcker and Heisenberg were inclined to read Bohr's views in the light of the Kantian philosophical tradition. As Weizsäcker was to put it later, after appreciating fully Bohr's philosophical position, 'as a physicist, I found it necessary to reflect on Kant. Space-time description corresponds to Kant's forms of intuition' (Weizsäcker, 1994, p. 171). The passage quoted above suggests this was Heisenberg's view as well.

It is worth briefly commenting on Weizsäcker's attitude to the relationship between Bohr and Kant, as it was to exercise a profound influence over Heisenberg. Indeed, through Bohr, Weizsäcker felt one could remain closer to the spirit of Kant's original insight than many other neo-Kantians had. Commenting on Ernst Cassirer's otherwise 'fine books on relativity and quantum theory' which appeared in the 1920s and 1930s, Weizsäcker was critical of Cassirer's capitulation to empiricist view that classical concepts may well become dispensable in physics. This line of thought, it seemed to Weizsäcker, 'robbed us of the valuable fruits of a well-fought battle' (Weizsäcker, 1994, p. 184). Indeed, others too saw Cassirer's work as marking a departure from traditional Kantian epistemology. Philipp Frank wrote that Cassirer's major

---

[3] My translation differs slightly from the one by Hayes (Heisenberg, 1952d).

work on quantum mechanics, *Determinism and Indeterminism in Modern Physics* 'is to be welcomed from the standpoint of logical empiricism as a highly successful attempt to continue the adjustment of traditional idealist philosophy to the progress of science' (Frank, 1975, pp. 184–5). Here Frank characterised Cassirer's view of physics as 'almost exactly that of logical empiricism' (Frank, 1975, p. 174). Weizsäcker felt that a more fruitful approach to the philosophical interpretation of quantum mechanics could be found by remaining true to Kant's original insight of a priori knowledge, albeit now modified through Bohr's notion of complementarity. As Weizsäcker succinctly put it: 'The alliance between Kantians and physicists was premature in Kant's time, and still is; in Bohr, we begin to perceive its possibility' (Weizsäcker, 1994, p. 185).

While there is little doubt that the discussions that took place in Leipzig between Heisenberg, Hermann and Weizsäcker helped to shape Heisenberg's own ideas, we should not infer from this that Heisenberg's neo-Kantianism was taken directly from Weizsäcker. Weizsäcker did exert a considerable influence on Heisenberg, as is evident from Heisenberg's discussion of metaphysical, dogmatic and practical realism in his work *Physics and Philosophy*. This is a mere paraphrasing of Weizsäcker's account of the same subject in his book, *The Worldview of Modern Physics* (Weizsäcker, 1952, pp. 107–13; Heisenberg, 1958d, pp. 75–7). Yet Heisenberg's reinterpretation of the Kantian a priori is different in certain important respects from Weizsäcker's, particularly in his emphasis on the constitutive dimension of human language, notably absent in Weizsäcker's account.

It has largely escaped the notice of scholars interested in the philosophy of quantum mechanics that Heisenberg devoted considerable attention to the philosophical problems of language from the 1930s onwards. The crucial chapter in his book, *Physics and Philosophy*, is entitled 'Language and Reality'. This theme is also found in his 1960 paper entitled '*Sprache und Wirklichkeit in der modernen Physik*' in which Heisenberg pointed out: 'It has long been supposed that the problem of language would play only a subordinate role in science. This is no longer true in modern physics' (Heisenberg, 1960b, p. 62). We find this view expressed in its clearest form in a private philosophical manuscript written by Heisenberg around 1942. The manuscript was later published in his *Collected Works* under the title, *Ordnung der Wirklichkeit*. This intriguing text, which has only recently attracted the attention of scholars interested in Heisenberg's thought, sheds important light on many of the more obscure passages in Heisenberg's later writings, where the language-reality problem is a recurring theme.

For Heisenberg, all knowledge of reality ultimately depends on the capacity of language to give form to what is given in experience. As Heisenberg was to put it in the 1942 Manuscript, 'every formulation of reality in language, not only grasps it, but also puts it into form and idealises it' (Heisenberg, 1984e, p. 289). Heisenberg's reflections on language share much in common with one of the dominant strands of the 'linguistic turn' in German philosophy in the twentieth century, according to which we must 'conceive of the world-making character of natural language in a strictly transcendental sense, that is, as constituting the world of possible objects of experience' (Habermas, 1999, p. 417). This view is evident in a crucial passage in the introduction to Heisenberg's 1942 manuscript:

> But definitely one must now and always realise that the reality of which we can speak is never reality "*an sich*", but is a reality of which we can have knowledge, in many cases a reality to which we ourselves have given form. If one objects to this last formulation that there is, nevertheless, definitively an objective world, completely independent of us and our thought ... which is what we truly aim for in the case of scientific research, then one must say, counter to this seemingly plausible objection, that the expression "there is" is nevertheless already derived from human language and cannot therefore really mean something which would not be connected, in one way or another, with our capacity for knowledge. For us, "there is" only the world in which the expression "there is" has a meaning.
>
> *(Heisenberg, 1984e, p. 236)*

This passage gives perhaps the clearest indication of the underlying Kantian impulse in Heisenberg's thought, and in doing so provides a deep insight into his philosophical viewpoint. Indeed it is through this that we can make sense of many of the passages that appear in Heisenberg's published writings throughout the 1950s. We cannot know reality in itself, but only reality as mediated through our language. Language is thus constitutive, not only of our thought and knowledge, but also of the possibility of an objective world. To this extent, Heisenberg claims that classical physics provides the language through which we can speak objectively about the world. One cannot dispense entirely with the concepts of classical physics, even in quantum mechanics, precisely because 'all the words which we use in physics in describing experiments, such as the words measurement or position or energy or temperature and so on, are based on classical physics and its idea of objectivity' (Heisenberg, 1989, p. 30).

Our language and thought is grounded in our basic intuitions of space and time. There is a similarity with Kant here, but as we shall see below, for Heisenberg, the forms of intuition do not originate in transcendental subjectivity, but emerge through our familiar dealings with the world around us. The language we use to describe events on the microscopic level, indeed the very structure of our thought, is intimately bound up with the spatio-temporal conceptualisation of the world. We extrapolate this from our familiar dealing with the world into a region very remote from ordinary experience. In the description of the interaction between the measuring instrument and the quantum-mechanical object, '*we are forced to use the language of classical physics*, simply because we have no other language in which to express the results' (Heisenberg, 1971, pp. 129–30, emphasis added). As Heisenberg explained in 1934:

> [T]he experimental questions which we ask of nature are always formulated with the help of intuitive concepts of classical concepts and more especially using the intuitive concepts of space and time. For indeed we possess no other language than the one adapted to the objects of our everyday environment with which we can describe for example the experimental set-up. Our experiences, too, can only be made in space and time.
>
> *(Heisenberg, 1934, p. 698)*

In *Physics and Philosophy*, Heisenberg again reinforced the point that we cannot avoid 'the subjective element in the description of atomic events, since the measuring device has been constructed by the observer, and we have to remember that *what we observe is not nature in itself, but nature exposed to our method of questioning*' (Heisenberg, 1958d, p. 57, emphasis added). We must recognise that in physics, our task is to pose 'questions about nature *in the language that we possess*' and 'to get an answer from experiment by the means at our disposal' (Heisenberg, 1958d, p. 57, emphasis added). This position is similar to the one we find in the preface to the second edition of the *Critique of Pure Reason*. There Kant asserted that reason does not approach nature 'in the character of a pupil who listens to everything the teacher has to say, but of an appointed judge who compels the witnesses to answer questions which he has himself formulated' (Kant, 1934, Bxii, p. 14). However, whereas for Kant the constitutive dimension of human knowledge originates in the immutable categories of reason, for Heisenberg it derives from language.

Our predisposition to use spatial concepts like 'position' and 'velocity' in the microscopic realm beyond what can be directly experienced is for Heisenberg, inextricably connected with our spatial form of intuition. In this context, Heisenberg frequently returned to the question of the a priori status of Euclidean space, time and causality in the light of modern physics. In his 1935 lecture, 'The Questions of Principle in Modern Physics', he argued that 'for an understanding of relativity theory it is essential to stress that the validity of Euclidean geometry is *presupposed* in the very instruments' we use in our experiments, in the measurement of physical space (Heisenberg, 1952a, p. 45). In the 1942 manuscript, he again referred to 'the question which has been continually discussed since Kant' concerning 'the *a priori* validity of the forms of intuition, space and time':

> These forms of intuition, such as they are given to us – with the validity of Euclidean geometry, the independence of space and time, etc. –, have proved themselves in man's dealings with the world, and they must owe their validity precisely to this

corroboration. Certainly they are to this extent *more* than merely an empirical given because they are, as Kant rightly emphasises, conversely the first presupposition of all possible experience.

(Heisenberg, 1984e, p. 284, emphasis in original)

It is important to note here that for Heisenberg, the a priori forms of intuition are not simply given in experience, nor do they originate from the structure of reason. Instead they are presuppositions for the possibility of experience, whose validity rests on our 'dealings with the world'. What becomes evident from Heisenberg's later discussion of the a priori in the context of the relation between quantum mechanics and Kant's philosophy is the articulation of a position, which, as Catherine Chevalley explains, avoids on the one hand the view 'that language has an *a priori* structure', and on the other 'that it is derived entirely from experience' (Chevalley, 1998, p. 163). As Heisenberg was to put it in his manuscript, this conception of space and time as forms of intuition 'corresponds therefore exactly to the intermediary between two extreme conceptions' one of which holds these forms to be 'absolutely valid, independent of all experience', and the other which holds them to be 'derived from experience' (Heisenberg, 1984e, p. 284). To this extent, our intuitions of space and time have what we might term a 'quasi-transcendental' status for Heisenberg, inasmuch as they arise out of our interplay with the world, at the same time form the presupposition for the description of experience.

7.4 Heisenberg's transformation of the a priori

It is important to distinguish two senses in which the a priori is no longer employed by Heisenberg in a strictly Kantian way in discussing quantum mechanics. The first of these can be stated as follows: while certain concepts such as those of classical physics can form the presupposition of all experience, they may be recognised to have a *limited range of applicability.* Heisenberg referred to this as 'the fundamental paradox of quantum theory' in his later writings. In quantum theory, we cannot ascribe to an electron a well-defined position or momentum at a given time. Yet such concepts are presupposed in the description of the electron as it manifests itself in its interaction with a measuring device. Though we speak of the uncertainty or indeterminacy in our concepts, we are compelled to employ the concepts of position and momentum. To this extent, such concepts are 'practically' indispensable. As Heisenberg puts it:

> The use of these concepts, including space, time and causality, is in fact the condition for observing atomic events and is, in this sense of the word, "*a priori*". What Kant had not foreseen was that these *a priori* concepts can be the conditions for science, and at the same time can have only a *limited range of applicability.* When we make an

experiment we have to *assume* a causal chain of events that leads from the atomic event through the apparatus, finally to the eye of the observer; if this causal chain was not assumed, nothing could be known about the atomic event. Still we must keep in mind that classical physics and causality have only a *limited range of applicability*. It was the fundamental paradox of quantum theory that could not be foreseen by Kant. Modern physics has changed Kant's statement about the possibility of synthetic judgments *a priori* from a metaphysical one into a practical one.

(Heisenberg, 1958d, p. 82, emphasis added)

Precisely because of the limited applicability of our classical concepts, an electron cannot be said to have a well-defined position and momentum. In quantum mechanics, we are forced to accept a practical, but not a strictly Kantian, interpretation of the a priori. The very concept of an object in space and time is only an *idealisation*. Yet, at the same time, such an idealisation is a necessary condition for the possibility of empirical knowledge. As he was to put it: 'Any concepts or words which have been formed in the past through the interplay between the world and ourselves are not really sharply defined' to the extent that 'we practically never know precisely the limits of their applicability'. This is 'true of even the simplest concepts like "existence" and "space and time"' (Heisenberg, 1958d, pp. 83–4). In the 1942 manuscript, Heisenberg draws particular attention to 'the unavoidable element of indeterminacy, this "suspense" [*Schwebende*] in language and thought' which confronts all knowledge of reality (Heisenberg, 1984e, p. 222).

We now turn to the second sense in which Heisenberg departs from the Kantian conception. Both Bohr and Heisenberg held the view that classical concepts remain necessary and indispensable in the description of experience, but their reasons for holding this view require further elucidation. In a lecture in 1931, Bohr stressed that 'we must realize the unambiguous interpretation of any measurement must essentially be framed in terms of the classical physical theories, and we may say that in this sense the language of Newton and Maxwell will remain the language of physicists for all time' (Bohr, 1931, p. 692). Writing to Bohr on 13 October 1935, Schrödinger cast doubt on the view 'that we *must* interpret observations in classical terms', and suggested that this might in fact be 'a temporary resignation', one that we may one day overcome (Bohr, 1996b, pp. 508–9). Bohr's reply unfortunately does not shed much light on the matter, as he simply restated what he saw as 'the seemingly obvious fact that the functioning of the measuring apparatus must be described in space and time' (Bohr, 1996b, pp. 511–12).

In his article examining Bohr's doctrine of classical concepts, Howard rejects the idea that Bohr's commitment to the indispensability of classical concepts stemmed from the view that we are somehow 'suspended in language'. As he puts it: 'To read Bohr as saying merely that we have to speak our mother tongue is to interpret the necessity of classical concepts as a contingent, historical necessity'. According to Howard, 'such a reading leaves the possibility that, as our language develops, we might outgrow this dependence' (Howard, 1994b, p. 209). Yet, as it turns out, Heisenberg seems to have adopted precisely this attitude.

Although on occasions he seems to have suggested that the concepts of classical physics will always remain the basis for objective science, at other times he expressed himself more cautiously. When pressed on this question by Kuhn, he conceded that 'from the historical point of view' it was not clear 'how people will talk about these problems in 1000 years from now' (AHQP, 27 February 1953, p. 25). Heisenberg here acknowledged that it is possible that in the future 'the language and the concepts [may] have changed so much that they would not use the Newtonian concepts at all anymore'. He argued that 'in order to describe phenomena, one needs a language. The language can only be taken from a *historical process*. Well, we do have a language and that is the situation in which we are' (AHQP, 27 February 1953, p. 26). Classical concepts are simply an extension or a refinement of the language that has adapted throughout human history to describe our everyday experience. Heisenberg's philosophical point of view has a similarity with Weizsäcker's analysis of Kant's concept of the a priori. To this end I will briefly examine the latter's writings on the subject.

In his book *The Worldview of Modern Physics*, originally published in German in 1949, Weizsäcker offered a penetrating critique of Kant's transcendental demonstration of a priori knowledge. Whereas for Kant, synthetic a priori judgements are necessary and universal, for Weizsäcker, the mere fact that we conceive of such judgements as necessary does not mean 'that we are also *justified* in conceiving' of them as such (Weizsäcker, 1952, p. 126). The concepts of classical physics can be regarded as 'necessary' only to the extent that we cannot conceive of how else to interpret our experience. This does not mean, however, that we are 'justified' in holding such concepts to be indispensable in any possible description of experience. Nevertheless, the concepts of space, time and causality may remain the a priori conditions of quantum mechanics, insofar as they remain the presuppositions of empirical inquiry:

It is in this sense that our interpretation of the concept "*a priori*" is intended. That every experiment is classically described we know no more certainly, than that every experience of the external world is spatial; in fact, the first proposition is the more questionable in the extent to which it asserts more. Both propositions have not logical, but factual necessity. We ought not to say, "Every experiment that is even possible *must* be classically described," but "Every actual experiment known to us *is* classically described, and we do not know how to proceed otherwise." This statement is not sufficient to prove that the proposition is *a priori* true for all, merely possible future knowledge; nor is this demanded by the concrete scientific situation. It is enough for us to know that it is *a priori* valid for quantum mechanics.

(Weizsäcker, 1952, p. 128, emphasis in original)

Though Weizsäcker never framed his discussion of the a priori in the context of the constitutive dimension of human language, the position he articulates in the above passage is close to the one adopted by Heisenberg. In the 1942 manuscript, Heisenberg argued that while space and time are conditions for the possibility of experience, 'the fact that we can have absolutely no experience other than in these forms of intuition – does not legitimate the supposition that the forms of intuition remain unchanged for all time' (Heisenberg, 1984e, p. 284). On some occasions, Heisenberg entertained the possibility that our forms of intuition may have even evolved biologically (Heisenberg, 1958d, p. 83; Heisenberg, 1984e, pp. 284–5).[4] Accordingly, our basic concepts of space, time and causality 'represent *for the time being* the final result of the development of human thought in the past, even in the very remote past; they may even be inherited and are in any case indispensable tools for doing scientific work *in our time*. In this sense they can be practically *a priori*' (Heisenberg, 1958d, p. 84, emphasis added).

Stimulated by Bohr's penetrating analysis of the problem of observation in the late 1920s and early 1930s, which revealed the essentially *classical* presuppositions underpinning all empirical knowledge, Heisenberg proposed a revision of Kant's concept of the *a priori*. Heisenberg's transformation of the a priori is significant for two reasons: first, because it provides the underlying philosophical framework for understanding Heisenberg's interpretation of quantum mechanics. And secondly, Heisenberg's own transformation of the a priori represents an original and largely unexamined approach to one of the most important themes of twentieth century philosophy. Michael Friedman and Don Howard have devoted considerable attention to examining the way in which the concept of the *a priori* was 'relativised' in the works of the logical empiricists, Reichenbach, Schlick, Carnap in the 1920s and 1930s (Friedman, 1994, 1996, Howard, 1994a). Heisenberg's own reconstruction of the a priori offers a new approach to the problem of knowledge left by Kant. Like the logical empiricists, Heisenberg rejected Kant's concept of the *a priori* as having the status of apodictic certainty. However, unlike them, he did not think of the a priori or non-empirical dimension of human knowledge as an arbitrary construction, nor as a set of hypotheses or conventions. Instead, for Heisenberg, it was grounded in the praxis of the historical development of human language and its refinement in classical physics. Indeed, his views appear to have more in

---

[4] Heisenberg was influenced in this view by the work of Konrad Lorenz (see Lorenz, 1977).

common with certain strands of the 'linguistic turn' in both analytic and continental philosophy (Habermas, 1999). We shall examine this in more detail in the next chapter.

Mara Beller has argued that Heisenberg's notions of the 'practical *a priori*' and the 'relative synthetic *a priori*' were nothing more than 'contradictions in terms' (Beller, 1999, p. 199). However, such a judgement fails to appreciate the way in which Heisenberg reinterprets Kant's concept of the *a priori*. Indeed, we can distinguish two distinct but interconnected ways in which this takes place. First, the a priori status of space, time and causality is retained by Heisenberg in the sense that they remain the presuppositions of experience, but the a priori is deprived of its necessity in Kant's sense. The indispensability of classical concepts originates from the historical fact that we have no other language through which we describe what is given to us in experience. Secondly, the concepts of classical physics are no longer understood to have universal applicability as they do for Kant. Beyond certain limits, the concept of the electron's location in space ceases to have applicability in quantum mechanics. In this way, the very notion of an 'object' in space and time is recognised to be an idealisation with limited applicability, but one which, for Heisenberg, ultimately remains indispensable for the possibility of human knowledge.

Although Heisenberg's philosophical discussions with Bohr, Weizsäcker and Hermann in the 1930s centred on the conditions for the possibility of knowledge, by the 1950s he began to turn his attention to the problem of reality in quantum mechanics. Heisenberg's later philosophical writings have sometimes been read as signifying a renunciation of the epistemological interpretation of quantum mechanics which was so characteristic of Bohr (Shimony, 1983, pp. 212–14). However, such accounts fail to see that Heisenberg's efforts to develop a quantum ontology can only be properly understood in the context of his epistemological viewpoint concerning the a priority of the classical concepts and the spatial and temporal forms of intuition. In this way, Heisenberg's thought, like Bohr's, remains fundamentally 'epistemological' in character, though of course they adopted different epistemological standpoints. A proper treatment of Heisenberg's concept of reality in quantum mechanics, and his much discussed concept of *potentia*, must therefore take into account his views on language. Heisenberg's turn to quantum ontology in the 1950s forms the subject of the next chapter.

8

The linguistic turn in Heisenberg's thought

By the 1940s, Heisenberg had arrived at the view that the problem of reality in quantum mechanics is inextricably intertwined with the constitutive dimension of language. This view marks the culmination of Heisenberg's philosophical interpretation of quantum mechanics. In this chapter, I will examine in more depth what I have termed the 'linguistic turn' in Heisenberg's later thinking, which was most clearly articulated in his unpublished philosophical manuscript written around 1942 and would continue to play a central role in his writings in the 1950s. Through these later texts we obtain a deeper understanding of the largely neglected language-reality problem in Heisenberg's philosophy of quantum mechanics, and how it shaped his views on the possibility of constructing a quantum ontology.

In the previous chapter, I argued that Heisenberg's later interpretation of quantum mechanics was underpinned by what could be termed a 'quasi-transcendental conception of language'. According to this view language is constitutive of the possibility of objective experience. In this chapter, I examine this view in more detail and situate it in historical context. Heisenberg's standpoint can be contrasted with the analytic conception of language taken up by the logical positivists, according to which meaning is ultimately derived from experience, and is closer in spirit to the 'linguistic turn' in German post-Kantian philosophy, which originated in the early nineteenth century with the work of Herder, Hamann and Humboldt, and was revived in the 1920s through the work of Ernst Cassirer. Critical here was Heisenberg's commitment to the world-disclosing function of language, though Heisenberg retained something of a realist perspective in recognising that language has a designative function, that is to say, language refers to a world which exists independently of how we conceive of it. The disjunction between the world-disclosing and designative functions of language in quantum mechanics would remain an unresolved tension in Heisenberg's later writings. The motion of a particle in quantum mechanics can be described by the Schrödinger wave equation, but this wave is not an 'objective reality'. This is because our concept of 'objective reality' derives its very *meaning* from the classical idealisation of things existing in space and time.

Heisenberg's introduction of the concept of *potentia* into quantum theory in the 1950s must be understood in the context of the language-reality problem. Yet Heisenberg's concept of *potentia* has often been interpreted as marking a foray into metaphysics, without paying sufficient attention to the epistemological and linguistic context in which it was proposed. The philosophy of language central to Heisenberg's later thought also provides an important clue to understanding his view of the much discussed measurement problem in quantum mechanics. The transition from potentiality to actuality, which occurs during the act of measurement, was not, for Heisenberg, a direct consequence of the conscious observer becoming aware of the result of the measurement as has sometimes been suggested. Rather, it emerges from our need to employ classical concepts in the description of objective experience. The 'actual', for Heisenberg, is not defined in terms of what is observable in an empiricist sense, but rather in terms of what can be grasped through the classical concepts of space and time. Understood in this way, the notion of potentiality signifies an attempt to go beyond the bounds circumscribed by classical language, but only to the extent that it represents the *possibility* of an 'objective reality' in space and time. In order to see this more clearly we must fist attempt to characterise Heisenberg's conception of language.

8.1 Heisenberg's conception of language

In Chapter 5, we saw that Heisenberg had become aware of a 'linguistic crisis' in quantum mechanics as early as 1926. This had been the problem that had initially prompted him to define concepts such as position and velocity operationally in 1927. After discussions with Bohr later that year, it slowly became apparent that an operational definition of concepts was not possible. By the late 1920s, Heisenberg would resign himself to the fact that in quantum mechanics we must simply accept that classical concepts are indispensable in the description of experience, in spite of the fact that they have only a limited range of applicability. Heisenberg would later describe this state of affairs as the 'fundamental paradox of quantum theory'. As he explained in an interview with Peat and Buckley: 'I will say that for us, that is for Bohr and myself, the most important step was to see that our language is not sufficient to describe the situation'. Yet, at the same time, Heisenberg and Bohr recognised that 'we must describe our experiments and results to other physicists' using classical concepts in spite of the fact that they 'have only a very limited range of applicability. That is a fundamental paradox which we have to confront. We cannot avoid it; we have simply to cope with it' (Peat & Buckley, 1996, pp. 6–7).

Heisenberg devoted considerable attention to the subject of language in his later writings. In *Der Teil und das Ganze*, Heisenberg recalled his conversations with Bohr in the 1930s, in which they had discussed language as the condition of the possibility of human thought: 'Language is, as it were, a net spread between people, a net in which our thoughts and knowledge are inextricably enmeshed' (Heisenberg, 1971, p. 138). In his interview with Peat and Buckley Heisenberg again stressed that in physics we must realise that 'we are bound up with a language; we are hanging in the language' (Peat & Buckley, 1996, p. 7). In the 1942 manuscript, Heisenberg argued that 'every formulation of reality in language, not only grasps it, but also puts it into form and idealises it' (Heisenberg, 1984e, p. 289). Emphasising the constitutive dimension of language and thought, Heisenberg pointed out that 'the reality of which we can speak is never reality "in itself", but is a reality of which we can have knowledge, in many cases a reality to which we ourselves have given form'. Whenever we attempt to describe a reality as it exists independently of ourselves, we must bear in mind that we are suspended in language and we 'cannot therefore really mean something which would not be connected, in one way or another, with our capacity for knowledge'. In perhaps his most striking statement of the world-disclosing function of language, Heisenberg declared: 'For us, "there is" only the world in which the expression "there is" has a meaning' (Heisenberg, 1984e, p. 236).

Heisenberg's reflections on the significance of language were in many respects similar to the view expressed by Aage Petersen, who was Bohr's assistant in Copenhagen in the 1950s. In the 1960s, Petersen suggested that the language-reality problem was at the heart of Bohr's philosophical viewpoint. We should treat Petersen's claim to accurately represent Bohr's views at this time with some suspicion. Petersen appears as but one of a line of thinkers all of whom claimed to give philosophical clarification to Bohr's own view. It should be pointed out that Bohr never wrote a treatise in any systematic way on his philosophy of language. His references to it are fragmentary, and what little is attributed to Bohr, comes from Petersen himself. In his exposition of Bohr's viewpoint, Peterson emphasised what he saw as the significant but often neglected aspect of Bohr's thought:

> Traditional philosophy has accustomed us to regard language as something secondary and reality something primary. Bohr considered this attitude toward the relation between language and reality inappropriate. When one said to him that it cannot be

language which is fundamental, but that it must be reality which, so to speak, lies beneath language, and of which language is a picture, he would reply, "We are suspended in language in such a way that we cannot say what is up and what is down. The word 'reality' is also a word, a word which we must learn to use correctly".

(Petersen, 1963, p. 11)

This conception of language presented here by Petersen was never fully developed in Bohr's published writings. Yet, according to Petersen, Bohr was convinced that 'It is wrong to think that the task of physics is to find out how nature *is*. Physics concerns what we can *say* about nature' (Petersen, 1963, p. 12). I suspect far too much has been made of remarks such as these in attempting to make sense of Bohr's philosophical viewpoint. It should be stressed that others who worked closely alongside Bohr at various stages, or who engaged in philosophical discussions with him, formed quite different impressions of his philosophy. Léon Rosenfeld, for example, who was Bohr's assistant in Copenhagen during the 1930s, argued that Bohr's complementary mode of thinking was 'the first example of a precise dialectical scheme' (Pauli, 1979, p. 481). Pascual Jordan and Philipp Frank, on the other hand, saw Bohr's views as signifying the triumph of positivism (Jordan, 1944, p. 159; Frank, 1975, p. 179). One finds a different version of Bohr's philosophy again in the writings of the Russian physicist Vladmir Fock, who defended complementarity as perfectly compatible with a version of Soviet Marxism after discussions with Bohr in 1957 (Fock, 1957, pp. 646–7). It is little wonder that over the past 30 years there has been a concerted effort on the part of scholars to disentangle Bohr's own views from those who professed to speak on his behalf (Howard, 1994b).

Whatever Bohr's own philosophical position, it seems clear that Heisenberg and Petersen expressed similar views regarding the relation between language and reality and its significance for the interpretation of quantum mechanics. It is important to realise that Heisenberg's view of language stands in sharp contrast to the analytical conception of language that had emerged in the works of Russell, Schlick and Carnap in the 1920s. According to this latter view, greatly influenced by Wittgenstein's *Tractatus logico-philosophicus*, language is made transparent by subjecting it to a rigorous logical analysis in which meaning is derived from experience. This was especially evident in the works of Russell and Carnap, for whom 'the method of explaining forms of thought by way of a logical analysis of language is still bound up with conventional empiricist epistemology' (Habermas, 1999, p. 425). One of the key aims of analytic philosophy in the first half of the twentieth century was to clarify language, in keeping with the project of Leibniz and Frege to construct a *characteristica universalis*. Heisenberg rejected such a revisionary project of analytic philosophy, first because, as we saw in the last chapter, for him language develops through a contingent, historical process, and secondly, because meaning is never fully determined. I will return to this latter point in due course.

By the 1930s, it seems Heisenberg had become quite critical of the Vienna Circle's analytic view of language. Writing to Schlick on 21 December 1932, Heisenberg described Schlick's view of the task of philosophy as 'completely off the track' (Mehra & Rechenberg, 2001, p. 691). His increasingly hostile reaction to positivism was also in evidence at the 'Unity of Science' congress in Copenhagen in 1936 (Strauss to Reichenbach, July 1936, ASP, Hans Reichenbach Collection [HR-013-35-07]). Reflecting on this position in a letter to Henry Stapp in 1972, Heisenberg acknowledged that the philosophical standpoint underpinning his own interpretation of quantum mechanics was much closer to 'the philosophy contained in the later papers of Wittgenstein' (in his post-analytic phase) than to 'the philosophy of the *Tractatus*' (Stapp, 1972, p. 1114).[1] In his interview with Peat and Buckley, Heisenberg expanded on this point:

> I would first like to state my own opinion about Wittgenstein's philosophy. I never could do too much with early Wittgenstein and the philosophy of the *Tractatus logico-philosophicus*, but I liked very much the later ideas of Wittgenstein and his philosophy of language. In the *Tractatus*, which I thought too narrow, he always thought that words have a well-defined meaning, but I think that is an illusion. Words have no well-defined meaning. We can sometimes by axioms give precise meaning to words, but still we never know how these precise words correspond to reality, whether they really fit or not. We cannot help the fundamental situation that words are meant as a connection between reality and ourselves – but we can never know how well these words fit reality. This can be seen in Wittgenstein's later work. I always found it strange when discussing such matters with Betrand Russell, that he held the opposite view.
>
> *(Peat & Buckley, 1996, p. 7)*

Heisenberg again expressed this same attitude to Wittgenstein's philosophy in a letter to Edward MacKinnon in 1974 (Chevalley, 1998, p. 185). Heisenberg then clearly saw his own position as somewhat at odds with the analytic philosophy of language. Whereas positivists like Schlick had maintained that 'the meaning of every proposition is exhaustively determined by its verification in the empirically given', Heisenberg categorically rejected such a view (Schlick, 1996, p. 58). Heisenberg's view has far more in common with another strand of German philosophy. Here, I am referring to the emergence of a transcendental, or perhaps more accurately, a quasi-transcendental conception

---

[1] Edward MacKinnon has argued that Bohr's view of language bears a similar relationship to the Anglo-American tradition in the twentieth century: 'In the mid-thirties Bohr was groping with the problems of the nature and grounds of meaning in ordinary language, problems that came to the forefront of the Anglo-American tradition only in the mid-fifties' (MacKinnon, 1984, p. 177).

of language in the twentieth century. The points of contrast and comparison between this conception and the analytic conception of language have been discussed by a number of scholars (Taylor, 1985; Habermas, 1999; Lafont, 1999). While his dialogue with Bohr was undoubtedly influential in shaping Heisenberg's view of language, his own 'linguistic turn' to a large extent parallels the 'linguistic turn' in German philosophy, which can be traced back to the early nineteenth century primarily with the work of Humboldt.

While Heisenberg's conception of language is by no means identical with, or as sophisticated as Humboldt's, they share two characteristic features. First, language is viewed as a constitutive principle of thought and knowledge. In this sense, it is regarded as the condition of both the possibility of the objectivity of experience and the intersubjectivity of communication. Secondly, the linguistic turn brings with it a dissolution of the sharp distinction Kant had drawn between a priori and a posteriori knowledge (examined in the previous chapter). Language then is both transcendental, in the sense that it is the condition for the possibility of knowledge, and at the same time empirical, because, unlike Kant's reason, language has of itself no universal and necessary forms (Lafont, 1994, p. 47; 1999, p. 1). Language for Heisenberg is not given to us a priori, but 'is formed from our continuous exchange with the outer world. We are a part of this world, and that we have a language is a primary fact of our life' (Peat & Buckley, 1996, p. 9).

The revival of Humboldt's philosophy of language in the twentieth century is sometimes associated with the hermeneutic philosophy of Heidegger and Gadamer, which first appeared in the late 1950s and early 1960s (Heidegger, 1971; Gadamer, 1989). However, we should not overlook the fact that Humboldt's philosophy rose to prominence in the 1920s through Ernst Cassirer's monumental three-volume work *The Philosophy of Symbolic Forms* (Cassirer, 1955). The first volume of Cassirer's work on language, published in 1923, illustrates the distinctly neo-Kantian dimension of Humboldt's philosophical view, according to which 'objective reality' is constituted through the form-giving activity of language:

> In this sense the objective is not the given but always remains to be achieved. Here Humboldt applies the Kantian critique to the philosophy of language. The metaphysical opposition between subjectivity and objectivity is replaced by their transcendental correlation. In Kant the object, as "object of experience", is not something outside of and apart from cognition; on the contrary it is only "made possible", determined and constituted by the categories of cognition. Similarly, the subjectivity of language no longer appears as a barrier that prevents us from apprehending objective being but rather as a means of forming, of "objectifying" sensory impressions.
>
> *(Cassirer, 1955, p. 158)*

Cassirer's account of the role of language in objectifying our sense experience is very close to Heisenberg's own conception of language. The view of the role of language expressed here stands in direct contrast with the commonly held instrumentalist conception, according to which language is merely the vehicle for expressing pre-linguistic thoughts. Humboldt's work remains close to traditional Kantian philosophy insofar as language plays an active role in the objectification of experience. While Heisenberg nowhere makes explicit a debt to Cassirer or Humboldt, certain sections of his 1942 philosophical manuscript seem to bear a striking resemblance to Cassirer's *Philosophy of Symbolic Forms* (Heisenberg, 1984e, pp. 279–94). In her detailed commentary on the manuscript, Catherine Chevalley has alluded to the parallels between Heisenberg and the work of Cassirer, and Humboldt's philosophy of language (Chevalley, 1998, pp. 149, 181, 232).

While it is not my aim here to argue that Heisenberg's view of language was directly influenced by the German philosophical tradition inaugurated by Hamann and Humboldt and continued by Cassirer, there is little doubt that it was closer in spirit to this tradition than to the analytic conception of language propounded by the likes of Carnap and Russell at the time. Whatever the source of inspiration for Heisenberg's conception of language, he was on several occasions explicitly critical of the view that language can be understood through logical analysis – a view most clearly expressed in works of the logical positivists. Heisenberg proposed that 'the grammars of different languages are quite distinct, and perhaps differences in grammar may produce differences in logic' (Heisenberg, 1971, p. 138). Here we find echoes of Humbolt's assertion of 'the primacy of grammar over logic' (Lafont, 1999, p. x). Heisenberg's later references to the 'dissolution of objective reality' in quantum mechanics should be read in the context of his quasi-transcendental conception of language. As we shall see, the language-reality problem thus forms the crux of his mature philosophy of quantum mechanics.

8.2 The language-reality problem in quantum mechanics

We are now in a position to examine in more detail the way in which this view of language underpins Heisenberg's later writings on quantum mechanics. In one sense he saw the discoveries of quantum mechanics as teaching us that the quantum world defies the classical mode of description, which has evolved to describe everyday experience. In the regions of nature that are not directly accessible to our senses, Heisenberg argued, 'we leave not only the realm of what can be directly experienced, but also the region in which our usual language formed itself and for which it is necessary' (Heisenberg, 1960b, p. 62). In describing such a world we are forced to use language, which is a human construction. For much of our experience our everyday language is adequate, but in the more remote realms of reality, such as the microscopic (atomic physics), or the macroscopic (general theory of relativity), the concepts of this language may no longer be applicable. In his 1953 lecture, *Das Naturbild der heutigen Physik*, which he presented at a conference organised by the Bavarian Academy of Fine Arts at Heidegger's request, Heisenberg drew attention to the radical change in the concept of reality that quantum mechanics had forced upon us:[2]

> Thus the objective reality of the elementary particles has been strangely dispersed not into the fog of some new ill-defined or still unexplained conception of reality, but into the transparent clarity of mathematics that no longer describes the behaviour of elementary particles, but only our knowledge of this behaviour. The atomic physicist has to resign himself to the fact that his science is but a link in the infinite chain of man's argument with nature, *and that it cannot simply speak of nature "in itself"*.
> *(Heisenberg, 1958b, p. 15, emphasis in original)*

The dissolution of 'objective reality' in quantum mechanics must not be misunderstood, as it has often been, as presenting a thoroughly subjectivist view which depends on the state of knowledge of the individual conscious observer. For Heisenberg, in regions where the space-time description breaks down – as in the case of quantum mechanics – we can no longer meaningfully speak of the electron as an 'objective reality'. The ψ-function of quantum mechanics denotes only the *probabilities* of certain states of affairs and as such quantum mechanics cannot describe nature as it exists 'in itself' but only our knowledge of nature. It is simply the case that all knowledge of objective reality must be grasped in the familiar concepts of space and time.

Throughout the 1950s, Heisenberg continued to emphasise that our conception of reality is ultimately dependent on the forms of thought and language we have at our disposal. In his article on the interpretation of quantum theory, written in 1955, he gave perhaps his clearest account of his own view of the Copenhagen interpretation of quantum mechanics, as a critical reaction against the dissenting voices of Einstein, Schrödinger, and Bohm:

> The criticism of the Copenhagen interpretation of the quantum theory rests quite generally on the anxiety that, with this interpretation, the concept of "objective

---

[2] The exchange of correspondence between Heisenberg and Heidegger leading up to the lecture is contained in the Werner Heisenberg Papers at the Max Planck Institute für Physik in Munich in folder 1953. See Heidegger to Heisenberg, 18 March 1953; Heidegger to Heisenberg, 9 June 1953; Heisenberg to Heidegger, 16 September 1953 (Carson, 1995, p. 88).

reality" which forms the basis of classical mechanics might be driven out of physics. As we have here exhaustively shown, this anxiety is groundless, since the "actual" plays the same decisive part in quantum theory as it does in classical physics. The Copenhagen interpretation is indeed based upon the existence of processes which can be simply described in terms of space and time, i.e. in terms of classical concepts, and which thus compose our "reality" in the proper sense.

He then went on to explain:

If we attempt to penetrate behind this reality into the details of atomic events, the contours of this "objectively real" world dissolve – not in the mist of a new and yet unclear idea of reality, but into the transparent clarity of a mathematics whose laws govern the possible and not the actual. It is of course not by chance that "objective reality" is limited to the realm of what Man can describe simply in terms of space and time. At this point we realize the simple fact that natural science is not Nature itself but a part of the relation between Man and Nature, and therefore dependent on Man. The idealistic argument that certain ideas are *a priori* ideas, i.e., in particular come before all natural science is here correct.

(Heisenberg, 1955, p. 28)

This passage gives one of the clearest expressions of Heisenberg's mature philosophy of quantum mechanics. By the 1950s, he would assert that our concept of physical reality 'is indeed based upon the existence of processes which can be simply described in terms of space and time'. Paradoxically, however, the investigation of the atom had revealed that 'the classical idea of "objectively real things" must here, to this extent, be abandoned' (Heisenberg, 1955, p. 27). In the atom, 'the contours of this "objectively real" world dissolve – not in the mist of a new and yet unclear idea of reality, but in the transparent clarity of a mathematics whose laws govern the possible and not the actual'. In this sense, we can only describe the quantum-mechanical 'object' as 'objectively real' to the extent that it can be described as existing in ordinary space and time. Although this is actually possible to a large extent under the conditions of experimental observation, even here there are limits, expressed by the uncertainty relations.

To understand more clearly the way in which Heisenberg identified 'objective reality' with the classical idea of things existing 'in' space and time, it is worth referring to Cassirer's discussion of the role of the concept of space in the constitution of objective reality. As he put it in the first volume of the *Philosophy of Symbolic Forms*, the Kantian critique of knowledge and the study of language show us that 'the act of spatial position and differentiation is the indispensable condition for the act of objectivization in general' (Cassirer, 1955, p. 203). Thus, whenever we speak of an 'object', we invariably invoke a language and a form of thought which presupposes that such an object exists somewhere in space:

The very *form* of spatial intuition itself bears within it a necessary reference to an objective *existence*, a reality "in" space … It is only when a content is determined in space, when it is distinguished by fixed boundaries from the undifferentiated totality of space, that it gains its own real form … And this logical fact is marked out in the construction of language, where the concrete designation of situation and space also serves as an instrument for defining the category of the "object".

(Cassirer, 1955, pp. 203–4)

This passage serves to show how the concept of an object is inextricably linked to the idea that it must exist in space. This is achieved through the construction of a language which is grounded in our form of spatial intuition. For Heisenberg, quantum mechanics did not signify the introduction of a radically new concept of 'objective reality', through *new* forms of intuition, but it had only made the existing concept a limiting concept, which can be approached but never fully realised in quantum mechanics. The task then is not to construct a new *quantum ontology*, but to adopt a fundamentally different epistemological *attitude* to the problem of reality. In grappling with the ontological problem posed by quantum mechanics, we are forced to realise that language figures in an activity which gives form to, and renders meaningful, the very concept of 'objective reality'.

8.3 The problem of meaning

While Heisenberg held the constitutive dimension of language to be fundamental, he did not entirely abandon the realistic intuitions that underpinned his thinking as a physicist. To this extent, he never doubted the existence of an external world, independent of thought and language. An underlying tension then emerges in Heisenberg's thought, which he confronted, but never fully resolved. On the one hand, language has a constitutive function, insofar as the objective knowledge is made possible only through language. On the other hand, language refers to a world independent of us and our thought. In other words, for Heisenberg, what we mean by our words or concepts is not determined by, and must be carefully distinguished from, what those words or concepts refer to. This distinction, which can be traced back to Frege's distinction between *Sinn* and *Bedeutung*, is crucial to the linguistic turn. As Charles Taylor explains, according to such a view, 'words are not just attached to referents like correlations we meet in nature; they are used to grasp these referents; that is they figure in an activity' (Taylor, 1985, p. 252).

Though Heisenberg never explicitly distinguished between meaning and reference in his published writings, the distinction is implicit in much of his later work.

Indeed it lies at the heart of the different understandings of the word 'definition' that appear in the works of Bohr and Heisenberg' papers in 1927, elaborated in Chapter 5. After his discussions with Bohr, Heisenberg came to the conclusion that the *meaning* attributed to such concepts as 'position' and 'velocity' derives from its general usage in classical physics, which in turn is nothing more than an effort to render more precise the concepts of everyday language. Yet, the *applicability* of such concepts to the electron is called into the question by quantum mechanics. We must therefore carefully distinguish between the two senses in which we may say our words are well defined. As Heisenberg explains:

> The words "position" and "velocity" of an electron, for instance, seemed perfectly well defined as to both their meaning and possible connections, and in fact they were clearly defined within the mathematical framework of Newtonian mechanics. But actually they were not well defined, as is seen from the relations of uncertainty. One may say that regarding their position in Newtonian mechanics they were well defined, but in their relation to nature they were not.
>
> *(Heisenberg, 1958d, p. 78)*

The passage quoted above is particularly important because it draws attention to the two different senses in which Heisenberg argued that our concepts can be 'well defined'. Although in physics, we may be able to clarify the meaning of such concepts as momentum, energy, force and mass through a series of interconnected definitions, we cannot know in advance the extent to which these terms really grasp hold of the 'external world'. In this context Heisenberg explained: 'Any concepts or words which have been formed in the past through the interplay between the world and ourselves are not really sharply defined' in the sense that 'we practically never know precisely the limits of their applicability'. Heisenberg asserts this to be 'true of even the simplest concepts like "existence" and "space and time"' (Heisenberg, 1958d, pp. 83–4). It is here we confront the metaphysical paradoxes of quantum mechanics.

Heisenberg's view of the language-reality problem, in particular the meaning–reference disjunction, though not fully developed, is closely linked to his thesis about the indeterminacy of meaning. Though Heisenberg recognised that only through language is the world disclosed to us, he insisted that the meanings of our words or concepts could never be made absolutely precise. There always remains a residual element of ambiguity. This theme emerges in Heisenberg's later writings and interviews. In the 1942 manuscript, he argued that the study of reality 'is thereby confronted at the outset by an abyss at the edge of which all human knowledge is played out. For, is it ever possible to express through language something that is entirely determined?' This question, Heisenberg explains is not to be understood as meaning that language fails to express what is clearly grasped in thought. Rather, it points to 'the unavoidable element of indeterminacy, this "suspense" [*Schwebende*] in language and thought' (Heisenberg, 1984e, p. 222).

The account offered here sheds light on Heisenberg's response to the question (which was the subject of vigorous debate between Bohr and Einstein in 1935) of whether or not a quantum-mechanical description of reality can be considered *complete*. As we saw in Chapter 6, Heisenberg argued that quantum mechanics can be considered complete in the sense that it is a closed theory which is not susceptible to further modification. To this extent, Heisenberg was convinced it was not possible to find a more complete theory of quantum mechanics. However, he did concede that quantum mechanics did *not* give a complete description of 'reality'. This is not because quantum mechanics fails, in Einstein's sense, to give a complete description of the 'true' state of motion of a particle. Rather, it is because, the very concept of 'objective reality' is meaningful only through the concepts of space and time. In this sense, we must 'realize that the incomplete knowledge of a system must be an essential part of every formulation in quantum theory' (Heisenberg, 1958a, p. 41). In *Physics and Philosophy*, Heisenberg defended the view that the description of reality in quantum mechanics must *necessarily* be incomplete:

> If therefore the atomic physicist is asked to give a description of what really happens in his experiments, the words "description" and "really" and "happens" can only refer to the concepts of daily life or of classical physics ... Therefore, any statement about what has "actually happened" is a statement in terms of the classical concepts and – because of ... the uncertainty relations – *by its very nature incomplete with respect to the details of the atomic events involved.*
>
> *(Heisenberg, 1958d, p. 127, emphasis added)*

Because we define the 'actual' in classical terms, Heisenberg argues that quantum mechanics forbids a complete knowledge of the actual. In his 1955 paper on the interpretation of quantum theory, Heisenberg wrote: '*Knowledge of the "actual" is thus*, from the point of view of the quantum theory, *by its nature always an incomplete knowledge*' (Heisenberg, 1955, pp. 27–8, emphasis in original). This view was shared by Hendrick Cassimir, who argued that 'the limited applicability of classical concepts to atomic and subatomic phenomena shows that the quantum mechanical description is incomplete' but we must realise that 'the quantum mechanical description is as complete as it can possibly be' given 'the limitations of our pictures of reality and of our language' (Casimir, 1986, p. 19). This limitation on knowledge, it must again be emphasised, is *not* a result of our ignorance of the electron's precise position and momentum. Rather, it arises from the paradoxical situation that on the one hand, according to quantum mechanics the electron does not move classically in space and time, while on the other hand our very notion of 'objective reality' is constituted through the concepts of space and time.

While the distinction drawn here between meaning and reference can help us make better sense of Heisenberg's defence of classical concepts despite the limits of their applicability, it should be recognised that this remained the source of much tension in his later writings. If 'objective reality' is defined for us through space and time, to what extent is it possible to speak of the *reality* of an external world independent of human thought and language? Christina Lafont spells this out in her account of the challenge for a philosophy which seeks to uphold the world-disclosing function of language. The trick then is 'to give an account of the *realist* intuitions highlighted by the linguistic function of designation ... without appealing to a metaphysical realism that would deny our unavoidably interpretative relationship to the world?' (Lafont, 1999, pp. xv–xvi). This characterises nicely the philosophical problem that confronted Heisenberg in his later years. When asked by Buckley whether there exists 'a fundamental level of reality' described by quantum mechanics, Heisenberg offered the following response:

> That is just the point; I do not know what the words *fundamental reality* mean. They are taken from our daily life situation where they have a good meaning, but when we use such terms we are usually extrapolating from our daily lives into an area very remote from it, where we cannot expect the words to have a meaning. This is perhaps one of the fundamental difficulties of philosophy: that our thinking hangs in the language.
>
> *(Peat & Buckley, 1996, p. 9)*

Here we find an unresolved tension in Heisenberg's thought brought about by the disjunction between the constitutive and referential functions of language in quantum mechanics. To this extent, in quantum mechanics, the very meaning of the words 'existence' and 'reality' becomes problematic. In the usual classical description, particles or waves 'exist' in space and time. However, in the case of a hydrogen atom, we cannot say an electron 'is' at a precise point in space at a given time. It is therefore not clear precisely what we mean when we say the electron *exists* in the atom. We can, of course, say that the electron is 'somewhere' in the vicinity of the atomic nucleus, but not precisely where. The very concept of 'objective reality', as Heisenberg repeatedly stressed, breaks down in the atom. Nevertheless, we feel compelled to say that the ψ-function, in describing the probability of finding an electron in a stationary state at a point in space or moving with certain momentum, refers to 'something'. Paradoxically, this something cannot be properly regarded as 'objectively real'. In the 1950s, Heisenberg introduced the concept of *potentia* to describe this strange kind of existence in quantum mechanics.

8.4 Actuality and potentiality

Undoubtedly, the most difficult problem confronting the interpretation of quantum mechanics is how to describe the inter-phenomenal or trans-phenomenal object. Reichenbach and Weizsäcker felt it necessary to discuss this problem in their epistemological and logical analyses of quantum mechanics (Reichenbach, 1944; Weizsäcker, 1955). As Heisenberg put it: 'It is of course tempting to say that the electron must have been somewhere between the two consecutive observations and that therefore the electron must have described some kind of path or orbit even if it may be impossible to know which path' (Heisenberg, 1958d, p. 49). Although the Copenhagen interpretation of quantum mechanics forbids any trajectory description of the electron's motion 'in between' observations, it is difficult to avoid the assumption that the electron must *exist*, in some form or another, when it does not interact with a measuring instrument.

Before examining in more detail Heisenberg's attitude to this dilemma, it is worth briefly drawing attention to the lack of clear consensus among scholars as to Bohr's view on this question. This we do for two reasons: as evidence of the elusiveness of the issues involved, and as a useful juxtaposition to Heisenberg's own view. Jan Faye claims that 'Bohr denied outright the intelligibility of ascribing reality to the transphenomenal object', and to this extent he should be interpreted as an anti-realist (Faye, 1991, p. 21). By contrast, Henry Folse argues that 'without presupposing *some* quantum mechanical object apart from the observed phenomena', Bohr held it to be 'quite meaningless to speak of the quantum descriptions of such phenomena as providing *complementary* evidence about the *same* object' (Folse, 1987, p. 162). Folse concludes, contra Faye, that Bohr should be understood as endorsing a realist position.

Bohr was extremely elusive on this issue. He seems to have been content to point to the semantic ambiguity that inevitably arises whenever we ask such a question. Indeed, his introduction of complementarity was motivated by the desire to avoid all such ambiguity. The physicist, Hendrick Casimir, recalled one occasion when Bohr was explaining the behaviour of an electron in a two-slit experiment in Copenhagen. In the discussion that followed, it was put to Bohr that 'the electron must be somewhere on its road from source to observation screen'. Bohr simply replied by posing the question: 'what is in this case the meaning of the word *to be*?' (Casimir, 1986, p. 17). Here again, the language-reality problem is brought to the fore. In a similar vein, Heisenberg referred what he saw as the inherent contradiction in demand to describe the electron between observations:

> The demand to "describe what happens" in the quantum-theoretical process between two successive observations is a contradiction *in adjecto*, since the word "describe"

refers to the use of classical concepts, while these concepts cannot be applied in the space between the observations; they can only be applied at the points of observation. It should be noticed at this point that the Copenhagen interpretation of quantum theory is in no way positivistic. For whereas positivism is based on the sensual perceptions of the observer as the elements of reality, the Copenhagen interpretation regards things and processes which are describable in terms of classical concepts, i.e., the actual, as the foundation of any physical interpretation.

(Heisenberg, 1958d, p. 127)

In this passage, Heisenberg is careful to emphasise that 'the Copenhagen interpretation of quantum theory is in no way positivistic' in that it does not restrict itself to what can be observed, but rather to what can be described in classical language. This marks a shift from the way he had expressed himself in 1925–7. Here we emphasise, it is not the 'observable' that is regarded as 'real', but rather the realm of what is 'classically describable'. Heisenberg does, however, acknowledge that we inevitably feel compelled to speak of the electron 'in itself'. In using this term, he is not suggesting we try to penetrate beyond a description of experience in terms of space and time – indeed to do so would be to reinterpret 'objective reality' in a radically different way – but rather the aim is simply to find a way of speaking about the electron *between* observations. The fundamental thesis that all 'points of experience' must be described in terms of classical concepts is left wholly intact by Heisenberg's effort to grapple with the dilemma of the 'inter-phenomenal' object. An appreciation of this point is critical for an analysis of Heisenberg's foray into quantum ontology and in particular his notion of *potentiality*.

In the passage quoted above, Heisenberg appears to forbid any ontological discussion of what happens to an electron between observations. Yet his realist inclinations led him to reflect further on the problem of the inter-phenomenal object in quantum mechanics. Here Heisenberg suggested that the concept of the 'state' in quantum mechanics could be of some use. The idea that the 'state' of a quantum-mechanical system represents the *probability* of an objective event occurring in space and time was regarded by Heisenberg as perhaps the most important step in the historical development of quantum theory. Heisenberg argued that we can always specify the state of a system through the Schrödinger ψ-function, or a state vector in Hilbert space, though this can represent only the state of our *knowledge* of the object; it does not represent an objective reality itself in space and time. However, by the mid-1950s, Heisenberg attempted to interpret this concept of the 'state' in quantum mechanics in ontological terms. As Weizsäcker was to put it, Heisenberg dared to express his conception of reality 'in a metaphysical imagery which Bohr's epistemological self-critique forbade him to use' (Weizsäcker, 1987, p. 290).

In his 1955 article on 'The Development of the Interpretation of Quantum Mechanics', Heisenberg first coined the term *potentia* in describing the 'state' of a system between observations. This term was self-consciously borrowed from Aristotle's philosophy (Heisenberg, 1955, p. 13) One finds the idea again in the 1955–6 Gifford lectures. However, there are anticipations of this idea earlier in the 1942 manuscript. There he states that quantum theory is 'that idealisation where reality [*Wirklichkeit*] appears at each instant as a determined abundance of possibilities for objective realisation [*Realisierung*]' (Heisenberg, 1984e, p. 256). Note the paradox here. Reality is defined as the possibility for objective realisation. In *Physics and Philosophy*, Heisenberg presented his most careful elaboration of this concept:

> This concept of "state" would then form a first definition concerning the ontology of quantum theory. One sees at once that this use of the word "state", especially the term "coexistent state", is so different from the usual materialist ontology that one may doubt whether one is using a convenient terminology. On the other hand, if one considers the word "state" as describing some potentiality rather than a reality – one may even simply replace the term "state" by "potentiality" – then the concept of coexistent potentialities is quite plausible, since one potentiality may involve or overlap with other potentialities.
>
> *(Heisenberg, 1958d, p. 159)*

This passage presents the clearest expression of Heisenberg's later attempt to develop a quantum ontology. It is critical to realise that Heisenberg does not define 'potentiality' as 'objective reality' because the concept of 'state' in quantum mechanics, unlike classical mechanics, represents only 'the possible not the actual'. Here Heisenberg turned to the Aristotelian concept of *potentia*. In a paper '*Sprache und Wirklichkeit in der modernen Physik*' published in 1960, he explained that 'in modern physics the concept of possibility, that played such a decisive role in Aristotle's philosophy, has moved again into a central place'. In this sense, 'the mathematical laws of quantum theory can be interpreted as a quantitative framing of this Aristotelian concept of that "*Dynamis*" or "*Potentia*" ' (Heisenberg, 1960b, p. 298). In his interview with Kuhn in 1963, Heisenberg explained that the 1924 Bohr–Kramers–Slater paper on virtual oscillators and the virtual radiation fields had first raised the possibility that a probability field could be interpreted not only as something in the minds of physicists, but as 'something in nature' (AHQP, 13 February 1963, p. 2).

But what exactly is this concept of *potentiality*? While Heisenberg claimed that the quantum-mechanical concept of probability has a parallel with the Aristotelian concept of *potentia*, scholars have been reluctant to place too great an emphasis on the analogy between Heisenberg's and Aristotle's concept. Both Shimony and Michael Carella are careful to point out that, unlike Aristotle's notion, Heisenberg's use of the concept of *potentia* in the context of quantum mechanics is devoid of all teleological significance (Carella, 1976, p. 34; Shimony, 1983, p. 212). Similarly, Patrick Heelan argues 'the philosophic setting of Heisenberg's thought is so entirely foreign to that of Aristotle that it is scarcely worthwhile to compare them' (Heelan, 1965, p. 152). Nevertheless, insofar as Heisenberg speaks of potentiality as a 'tendency', there does appear to be some link between his use of the term and Aristotle's. Shimony concludes that 'Heisenberg has drawn from quantum mechanics a profound and radical metaphysical thesis: that the state of a physical object is [nothing but] a collection of potentialities' (Shimony, 1983, pp. 214–15). Indeed, this has often been viewed as the defining characteristic of his later philosophy of quantum mechanics (Northrop, 1958, pp. 13–14). However, these attempts fail to recognise that Heisenberg's notion of potentiality must be understood within the context of his philosophy of language. If 'objectively reality' or 'actuality' is constituted through the space-time description, then potentiality refers to that which can be 'actualised' – that is to say what can become 'objectively real' in this sense. It is only because under certain experimental conditions an electron can be 'actualised' in space and time that we can speak of it as a 'potentiality'.

We can gain a deeper insight into the way the concept of *potentia* fits into Heisenberg broader philosophical view by examining his idea of 'actualisation' through measurement. As Shimony puts it: 'how does the transition from potentiality to actuality take place? In other words how does the reduction of the wave packet occur?' (Shimony, 1983, p. 213). It is sometimes thought that Heisenberg's account of the reduction of the wave packet is brought about by the act of recognition by the conscious observer. Howard, for example, has argued that Heisenberg accepted as 'distinctly subjectivist of the role of the observer' in quantum mechanics (Howard, 2004, p. 677). Though this idea has often been associated with the Copenhagen interpretation, so far as I can tell, none of the founders of quantum mechanics, with the possible exception of von Neumann, ever seriously entertained this notion in the 1930s. Indeed, as we shall see, Heisenberg categorically rejected such an interpretation.

The idea that the conscious observer plays a crucial role in measurement in quantum mechanics is perhaps first expressed in von Neumann's *Mathematische Grundlagen der Quantenmechanik*, published in 1932 and was developed further in London and Bauer's 1939 essay *La Théorie de l'Observation en Méchanique Quantique*. The essay was an extensive analysis of von Neumann's theory of measurement, in which the measuring instrument, contra Bohr, is treated quantum-mechanically. Here the authors emphasised 'the essential role played by the *consciousness of the observer*' in the collapse of the wave packet which occurs during the act of measurement (London & Bauer,

1983, p. 251). In the early 1960s, Wigner defended the von Neumann–London–Bauer view of measurement as 'the orthodox view' (Wigner, 1983a,b). Yet, as Shimony has rightly pointed out, 'For Wigner the term "orthodox" refers not to Bohr's Copenhagen Interpretation and its variants but to von Neumann's' (Shimony, 1997, p. 407). Indeed, Heisenberg was careful to distance himself from this subjectivist interpretation of quantum mechanics. He would spell this out in *Physics and Philosophy*:

> [T]he transition from the "possible" to the "actual" takes place during the act of observation ... *It applies to the physical, not the psychical act of observation.* And we may say the transition from the "possible" to the "actual" takes place as soon as the interaction of the object with the measuring device, and thereby with the rest of the world, has come into play; it is not connected with the act of registration of the result, by the mind of observer. The discontinuous change in the probability function, however, takes place with the act of registration, because it is the discontinuous change of our knowledge in the instant of registration that has its image in the discontinuous change of probability function.
>
> *(Heisenberg, 1958d, p. 54, emphasis added)*

This passage has not always been given the careful consideration it deserves. While Heisenberg readily concedes that the discontinuous change in the ψ-function, as a representation of the state of our *knowledge* of the system, takes place with the registration in the mind of the observer, he explicitly denies that the transition from the 'possible' to the 'actual' should be understood as entailing a radically subjectivist view of measurement. Heisenberg was categorical on this point: 'quantum theory does not contain genuine subjective features, it does not introduce the mind of the physicist as part of the atomic event' (Heisenberg, 1958d, p. 55). As he explained in 1935, we must 'start from the assumption that the appearance of a definite event is an objective fact that *does not depend on our observation of this event*' (Heisenberg, 1985b, pp. 414–15, emphasis added). Though not immediately obvious from the passage quoted above, Heisenberg's account of the transition from the 'possible' to the 'actual' in the act of measurement depends on his understanding of the language-reality problem in quantum mechanics.

According to Heisenberg, once an observation is made, for example through a mark left on a photographic plate, we must assume that an 'interaction' has taken place between the object and the instrument. This is an assumption we cannot avoid if the concept of observation is to have any meaning at all. However, as discussed in the previous chapter, such an interaction cannot be described in quantum-mechanical terms. In order for the observation to count as an observation, we must assume that the interaction has definitely taken place somewhere in space and time, and to this extent, it has already become a part of 'objective reality'. The mark on the photographic plate indicates that something has 'happened'. As Heisenberg explained: 'the formalism of quantum theory does not as a rule lead to a definite result; it will not lead, e.g., to the blackening of the photographic plate at a given point'. The Schrödinger equation gives only a dynamical law for the *probability* that such occurrence will take place. 'It is the "factual" character of an event describable in terms of the concepts of daily life which is not without further comment contained in the mathematical formalism of quantum theory'. To this extent, 'the transition from the "possible" to the "actual" is absolutely necessary here and it cannot be omitted from the interpretation of quantum theory' (Heisenberg, 1958d, p. 121). It is not that the interaction of object and instrument *brings about* the actuality of the electron from a state of potentiality in a causal sense. Rather, it is simply that we must assume that the electron is 'objectively real' (i.e. localised in space and time) in order to account for the 'fact' that it has interacted with the instrument.

The 'transition from the possible to the actual' is therefore completely misunderstood if it is interpreted as a collapse of a physically real extended wave-packet in space. Nor should it be interpreted in Berkeley's sense: *esse est percipi*. Rather, we must understand the 'actual' and the 'possible' as two modes of description, both of which employ the language of space and time at some level. The transition from potentiality (a quantum-mechanical description) to actuality (a classical space-time description) must be understood as a transition from one *mode of description* to another. The two modes of description – the possible and the actual – are deemed complementary in Heisenberg's, though not in Bohr's, sense of the term.

In Heisenberg's view, we are forced to speak in terms of the actual and the possible in quantum mechanics, because ultimately 'we are suspended in language'. If it were possible to transcend the language of objects in space and time in the description of physical experience or empirical reality, it might be possible to bridge the chasm between actuality and potentiality. However, as Heisenberg tells us: 'There is no use in discussing what could be done if we were other beings than we are' (Heisenberg, 1958d, p. 55). We must be content with the fact that an electron in itself cannot be regarded as 'objectively real' in any ordinary sense of the term:

[I]f one wishes to speak about the atomic particles themselves one must either use the mathematical scheme as the only supplement to natural language or one must combine it with a language that makes use of a modified logic or no well-defined logic at all. In the experiments about atomic events we have to do with things and facts, with phenomena that are *just as real* as any phenomena in daily life. But the atoms or the elementary particles themselves *are not as real*; they form a world of *potentialities* or *possibilities* rather than one of things or facts.

(Heisenberg, 1958d, p. 160, emphasis added)

In speaking of 'modified logic' Heisenberg is here referring to the attempts of thinkers like Reichenbach and Weizsäcker who attempted to replace the traditional two-valued logic of Aristotle with a more complex logical system. Such attempts to extend logic were not investigated by Heisenberg in any systematic fashion. He was content on most occasions simply to speak of the electron itself as a 'potentiality'. The passage above does, however, illustrate nicely one of the inherent difficulties that plagued discussion of the physical reality of atoms and electrons, from the inception of quantum mechanics. In Heisenberg's later philosophy, the concept of reality was inextricably linked to his view of the constitutive dimension of human language, and to the problems, characteristic of the linguistic turn more generally, associated with the distinction between meaning and reference.

Heisenberg's later writings on quantum mechanics reflect his concerns with both the constitutive dimension and the limitations of human language. Here we are confronted with the problem that 'objective reality' is ultimately constituted through classical language, whereas quantum mechanics can describe only the *possibility* of finding an object existing in space and time. To this extent, we are no longer able to describe the electron itself as fully real. This paradox, resulting from the disjunction between the classical *meaning* we attach to the concept 'objective reality' and the fact that the ψ-function must *refer* to something, lies at the heart of Heisenberg's attempt to overcome the language-reality problem in quantum mechanics. The division between actuality and potentiality must be understood within the context of a continuing commitment to the quasi-transcendental conception of language. This is significant, not only in presenting a different, and previously unexplored dimension of Heisenberg's philosophy of quantum mechanics, but also in situating him in the context of the German philosophical tradition not normally associated with modern physics. A proper appreciation of the philosophical depth of Heisenberg's thought requires that we attend carefully to his reflections on the related problems of meaning and language, which, as we have seen, acquired particular salience by the early 1940s.

Conclusion

Heisenberg's interpretation of quantum mechanics gradually came to centre itself on a philosophy of language. It is perhaps fitting then that his conversations and discussions with those around him were critical to the way his thought unfolded. His work *Die Teil und das Ganze*, written in the form of a Platonic dialogue, is a testament to the value Heisenberg placed on the dialogical form of thinking. Indeed Weizsäcker described the work as 'the only real Platonic dialogue in contemporary literature that I know' (Weizsäcker, 1988, p. 197). While Heisenberg's reconstruction of past conversations and events have been treated with some suspicion by historians, particularly in the light of the recent literature on his political involvement in the atomic bomb project in Nazi Germany during the war years, the dialogical form which Heisenberg chose tells us something important about his style of thinking. The protagonists of Heisenberg's work are real, and the conversations are historical reconstructions of actual conversations that took place throughout Heisenberg's lifetime. Heisenberg's philosophy, perhaps even more so than his physics, was shaped not simply through his own private meditation, but through his dialogue with other physicists and philosophers. Undoubtedly, the most significant interlocutor for Heisenberg was Bohr, but his discussions with Pauli, Einstein, Schrödinger, Weizsäcker, Hermann, and Schlick all left their mark on Heisenberg's thought.

Through discussions with other philosophers and philosophically minded physicists Heisenberg was brought into contact with many of the important philosophical currents of his time. Chevalley, in her introduction to the 1942 manuscript, suggests that Heisenberg's philosophy of quantum mechanics has not been fully understood or appreciated 'because of the failure to place it in its proper context' (Chevalley, 1998, p. 138). By situating Heisenberg's thought against the background of post-Kantian philosophy in the German-speaking world, it becomes evident that many of the themes with which he was concerned

172

in the 1920s and 1930s were 'consistent with major issues in philosophical thought at the turn of the twentieth century' (Chevalley, 1994, p. 51). Heisenberg's discussion of observability, visualisability, the problem of meaning, the a priori–a posteriori distinction, and the world-disclosing function of language, fit neatly with the wider philosophical setting of the first third of the twentieth century. Heisenberg was willing to discuss the philosophical questions posed by quantum physics with thinkers as different as Schlick and Heidegger. To this extent, Heisenberg's thought was not exclusively shaped by the rise of logical positivism in the 1930s, as has sometimes been contended, but reflects the influence of a number of different strands of thought in the German-speaking world during this time.

The central thesis in this book is that Heisenberg's thought moved away from an early preoccupation with positivism in the mid-1920s, towards a philosophy of language, which by the 1940s would hold that the Kantian forms of intuition and the space-time description are, paradoxically, both the condition for the possibility of objective experience, and yet at the same time, only of limited applicability in the quantum world. From as early as 1924, Heisenberg became aware that quantum theory demanded a renunciation of the foundations of classical physics. He was by this time convinced that the electron was not in reality a point-mass that moved in a well-defined orbit around the atom. Nor did Heisenberg think the electron was a wave spread out in the space surrounding the atomic nucleus. It therefore seemed clear to Heisenberg that quantum mechanics would require a new epistemology and ontology.

Between 1925 and 1927, Heisenberg appealed to a number of different strands of positivist thought in an attempt to construct a quantum mechanics which dispensed with the classical concept of a particle moving in a continuous trajectory in space and time. This is evident in three different ways: first, in the renunciation of the electron orbit in 1925, which he defended by recourse to the principle of observability; secondly, in the rejection of the classical ideal of understanding as entailing visualisation in space and time in favour of an instrumentalist conception of 'intuitive understanding'; and thirdly, in the operational definition of concepts, which would replace the classical concepts of position and velocity. All these efforts indicate that the empiricism to which Heisenberg appealed directly was closely and explicitly associated with Einstein's theory of relativity. The principle of observability had risen to prominence through Einstein's critique of the notions of absolute space and time. The instrumentalist conception of 'intuitive understanding' had found its clearest illustration through Einstein's conception of a non-Euclidean space. The operational definition of concepts to which Heisenberg appealed in his 1927 paper had underpinned Einstein's concept of simultaneity in the special theory of relativity. It therefore comes as no surprise to find that Heisenberg later admitted that he had been 'impressed by Einstein's way of doing things' (AHQP, 30 November 1962, p. 4).

However, by the early 1930s, the positivism characteristic of Heisenberg's early papers on quantum mechanics would be replaced by a different philosophical viewpoint. This transformation in Heisenberg's outlook can be largely attributed to Bohr's influence, although other physicists and philosophers were undoubtedly important to his shift. Einstein himself, whose own philosophy had by the 1920s moved away from positivism, was instrumental in persuading Heisenberg that the idea that only observable quantities should be admitted into physical theories was problematic. The concept of observation would continue to be an important theme in Heisenberg's philosophical thought, again becoming the subject of discussion in 1927, this time with Bohr. Heisenberg's introduction of the gamma-ray microscope thought-experiment in his 1927 paper on the uncertainty relations had provoked Bohr's criticism. While Heisenberg had introduced this imaginary experiment to provide an operational definition of position, Bohr rejected the verification theory of meaning, and insisted that classical concepts remain indispensable in the description of experience, despite the inherent limitations on their applicability. This would prove to be the decisive turning point in Heisenberg's interpretation of quantum mechanics.

After exhaustive discussions with Bohr in 1927, Heisenberg abandoned his earlier view that it was possible and indeed necessary to find new 'quantum' concepts of space and time or an operational definition of kinematic concepts. He now resigned himself to the view that quantum mechanics would rest on the indispensability of classical concepts like position and momentum. In the 1930s, influenced by his discussions with Weizsäcker and Hermann in Leipzig, Heisenberg turned his attention to the epistemological foundations for this view. Here he proposed, along with Weizsäcker, a pragmatic transformation of Kant's notion of the a priori. Heisenberg argued that concepts such as space, time and causality can be regarded as 'practically *a priori*', insofar as they remain the conditions for the possibility of experience and even of 'objective reality', though they are not universal and necessary in a strictly Kantian sense. We must use classical concepts in the description of experiments in quantum theory, despite the fact that there are limits to their applicability. Such concepts are, for Heisenberg, historically contingent insofar as they are 'formed from our continuous exchange with the outer world' (Peat & Buckley, 1996, p. 9). Yet they remain indispensable in our time, because we have no other language through which we can constitute 'objectively reality'. In this sense, language for Heisenberg is accorded a quasi-transcendental status: 'every formulation of reality in language, not only grasps it, but also puts it into form and idealises it' (Heisenberg, 1984e, p. 289). Our very notion of 'objective reality' in space and time is thus an idealisation with limited applicability, but one which, for Heisenberg, ultimately remains indispensable for the possibility of human knowledge.

Heisenberg's later formulation of the concept of *potentia* to describe the inter-phenomenal object in quantum mechanics must be understood in the context of his linguistic turn. If 'objectively reality' or 'actuality' is constituted through the space-time description, then potentiality refers to what can be 'actualised' in an experimental set-up. The electron, as it were, only becomes 'objectively real' during the act of observation. However, this transition from the potential to the actual is not to be understood as taking place because it is apprehended by a conscious observer. Rather, it must be understood as a consequence of the need to shift from one mode of description to another. The distinction between 'actuality' and 'potentiality' in quantum mechanics, and the paradoxical nature of the quantum world is, for Heisenberg, ultimately a consequence of the fact that all thinking about physical reality is suspended in a language which is essentially classical.

Despite Heisenberg's protestations in later years that he was not a positivist, this label has stuck. This is largely because of the way in which later writers have construed the controversy over the interpretation of quantum mechanics in the 1930s. As Catherine Chevalley explains: 'The majority of usual expositions of the divergence of the interpretations of the quantum theory translate this divergence in terms of a conflict between two positions, that of positivism and that of realism' (Chevalley, 1992, p. 67). Indeed, in 1930, Hans Reichenbach declared that Heisenberg's work in quantum mechanics had provided 'a confirmation of those previous deliberations which introduced a positivist-empiricist element into classical physics' (*Diskussion über Kausalität und Quantenmechanik*, 1931, p. 188). Similarly, Pascual Jordan argued that positivism is 'the epistemological viewpoint of Bohr and Heisenberg' (Jordan, 1936, p. vii). As Euan Squires points out: 'The Copenhagen interpretation and the prevailing fashion in philosophy, which inclined to logical positivism, were mutually supportive' (Squires, 1994, p. 118). The historical reconstructions of the positivism of the Copenhagen school and the realism of physicists like Einstein, de Broglie and Schrödinger, has only served to present 'a simplified description of the situation, to the detriment of the divergences which existed among the founders of quantum mechanics … and among their opponents' (Chevalley, 1992, pp. 67–8). Recent scholarship has shown that neither Einstein nor Schrödinger should be considered realists in any straightforward sense of the term (Fine, 1986; Götsch, 1992; Bitbol, 1996). Moreover, as Anton Degen rightly points out, 'most of the founders of quantum mechanics did not subscribe to a positivist philosophy of science, even though they may have had leanings in that direction earlier in their careers' (Degen, 1989, p. 17). This is particularly evident in the case of physicists like Pauli and Heisenberg.

Here we might ask whether the later Heisenberg was a realist or an anti-realist. Certain passages in Heisenberg's later writings certainly appear to place him squarely in the anti-realist camp. In his discussion of the constitutive dimension of human language, influenced by his reading of Kant, Heisenberg argued that 'the reality of which we can speak is never reality "in itself", but is just a reality of which we can have knowledge, in many cases a reality to which we ourselves have given form' (Heisenberg, 1984e, p. 236). Yet, other passages make it clear that Heisenberg was a realist of sorts. As he was to put it in *Der Teil und das Ganze*: 'If nature leads us to mathematical forms of great simplicity and beauty ... we cannot help thinking that they are "true", that they reveal a genuine feature of nature'. When we do discover such laws, which have a simple axiomatic structure and at the same time describe a wide range of phenomena, we are led to conclude that 'they must be part of reality itself, not just our thoughts about reality' (Heisenberg, 1971, p. 68). Some authors, most notably Mara Beller, have simply accused Heisenberg of downright inconsistency here (Beller, 1999, p. 172). However, this apparent contradiction can be resolved by a careful reading of Heisenberg's texts. As I argued in Chapter 3, Heisenberg's realism was in many respects close to a position we would today identify as *structural realism*. In the case of 'closed theories' like quantum mechanics, the mathematical laws capture some essential feature of the mathematical *structure* of reality but they cannot tell us the nature of this underlying reality. In our attempts to grasp the essence of this underlying reality, we are immediately 'confronted at the outset by an abyss at the edge of which all human knowledge is played out' (Heisenberg, 1984e, p. 222). As he put it in poetic terms: 'The human capacity to understand is unlimited. Of the ultimate things we cannot speak' (Heisenberg, 1984e, p. 226).

The relationship between Heisenberg's and Bohr's philosophies of quantum mechanics is one of the central themes of this book. In much of the existing literature which examines the Bohr–Heisenberg dialogue, Bohr is usually represented as the central figure while Heisenberg the peripheral. This book inverts this image, by making Heisenberg the central focus. Patrick Heelan has argued that Heisenberg's philosophy 'except for a short period when he closely collaborated with Bohr, was very different and became increasingly so with the passage of time' (Heelan, 1965, pp. xii–xiv). Yet, Heelan's study fails to recognise that Bohr probably remained the single most important *philosophical* influence on Heisenberg. Through his discussions with Bohr, Heisenberg became convinced of the epistemological primacy of 'classical language' in the physical interpretation of quantum mechanics. While it is certainly true that Heisenberg's understanding of complementarity differed from that of Bohr, the latter's underlying philosophical standpoint, in particular his epistemological analysis of the observation problem in quantum mechanics in the late 1920s, would exert a decisive influence on Heisenberg.

Notwithstanding Heisenberg's intellectual debt to Bohr, he maintained a certain independence from his teacher, which others have acknowledged but inadequately explored. It is generally accepted that after a brief period of disagreement in 1927, the two men reached agreement on the meaning of the indeterminacy principle, wave–particle duality and the significance of complementarity. However, a careful reading shows that Heisenberg and Bohr gave different meanings to terms like 'complementarity' and 'wave–particle duality'. While we may accept MacKinnon's characterisation of the Copenhagen interpretation as including 'the idea that photons, electrons and other "particles" exhibit both wave and particle properties', Bohr and Heisenberg interpreted this idea in very different ways (MacKinnon, 1985, p. 110). Although Heisenberg publicly defended Bohr's idea of complementarity, as we have seen, his own interpretation of the doctrine differed in several crucial respects. It is not only in the divergent meanings each ascribed to wave–particle duality and complementarity that we find disagreement between Heisenberg and Bohr. They also held fundamentally different attitudes to the ambiguity in our use of classical concepts in quantum mechanics. Whereas Bohr strove for a precise way of speaking about the quantum world through complementarity, Heisenberg remained resigned to the view that in quantum mechanics our descriptive language cannot be made free from ambiguity. Heisenberg was far more inclined than Bohr to ascribe philosophical significance to the mathematical formalism of quantum mechanics. In disentangling their respective views this book sheds new light on the historical myth of the Copenhagen interpretation, much discussed in the recent works of Hendry, Beller, Chevalley and Howard (Hendry, 1984; Beller, 1999; Chevalley, 1999; Howard, 2004).

The divergence between Heisenberg and Bohr is perhaps best illustrated in their respective attitudes to the completeness of quantum mechanics. Throughout the 1930s, Bohr famously defended the 'completeness' of the theory against the attacks from Einstein on the basis of his notion of complementarity. Yet, in the mid-1930s, Heisenberg also defended quantum mechanics as 'complete', but unlike Bohr he did so from the perspective that it was a 'closed theory' with a well-defined axiomatic structure which accounted for a wide range of experimental phenomena. To this extent, Heisenberg argued, quantum mechanics had reached a state of completion and was no longer susceptible to further modification. Heisenberg presented his most elaborate defence of the completeness of quantum mechanics in his unpublished reply to the EPR paper in 1935. There, he employed the 'cut' argument, insisting that the position of the diving line or 'cut' between the 'quantum-mechanical' object and the 'classical' measuring instrument cannot be established physically, but is subject to choice. However, in the 1950s, Heisenberg also argued that quantum mechanics must also be regarded as an *incomplete* description of physical reality. This is because on the one hand, 'objective reality' is limited to the realm of what can be described simply in terms of space and time, while on the other hand, the laws of quantum mechanics 'govern the possible and not the actual'. In this sense, our knowledge of reality is 'from the point of view of the quantum theory, *by its nature always an incomplete knowledge*' (Heisenberg, 1955, pp. 27–8).

Recent scholarship into Bohr's philosophy of physics has revealed that his views continued to evolve and change throughout the 1930s. As this book shows, the same is true of Heisenberg. Although Heisenberg would later claim that from 'the spring of 1927 one has a consistent interpretation of quantum theory, which is frequently called the "Copenhagen interpretation"' (Heisenberg, 1958d, p. 44), his *philosophical* viewpoint continued to unfold after this point, partly as a result of his philosophical discussions with Bohr and Weizsäcker. Far from bringing Heisenberg's reflections on quantum mechanics to finality, the dialogue with Bohr in 1927 represented a crucial turning point. His earlier empiricist tendencies began to make way for a different philosophical outlook rooted in the epistemological primacy of classical language. By the 1950s, Heisenberg's philosophy of quantum mechanics had centred on the language-reality problem. Here he saw the world-disclosing and designative aspects of human language as inextricably connected to the problem of reality in quantum mechanics.

This preoccupation with language in Heisenberg's later thought has gone largely unnoticed by historians and philosophers of science. Indeed, notwithstanding the work of Edward MacKinnon, the philosophy of language has played little role in the debate over the interpretation of quantum mechanics. Traditionally, the battle lines of this debate have been drawn between realism and anti-realism. It is worth highlighting that Heisenberg's mature philosophy, and in particular his linguistic turn, has much in common with one of the dominant themes of twentieth-century philosophy (though as I have argued, Heisenberg's views have more in common with the continental than the analytic tradition). By the middle of the century, Heisenberg's writings would offer a new insight into the 'problem of reality' by inviting his readers to reconsider the philosophical challenge of quantum mechanics in the light of the 'linguistic turn'. Few of his contemporaries were able to grasp the significance of what Heisenberg had to say. More than half a century later we are perhaps better placed to understand his philosophical insight into the quantum revolution.

References

Archives for the History of Quantum Physics. (1986). 301 microfilm reels. Philadelphia: American Philosophical Society.

Archive for Scientific Philosophy in the Twentieth Century. Pittsburgh: University of Pittsburgh.

Beller, M. (1983). Matrix theory before Schrödinger. *Isis*, **74**, 469–91.

Beller, M. (1985). Pascual Jordan's influence on the discovery of Heisenberg's indeterminacy principle. *Archive for the History of Exact Sciences*, **33**, 337–49.

Beller, M. (1988). Experimental accuracy, operationalism, and the limits of knowledge – 1925 to 1935. *Science in Context*, **2**, 147–62.

Beller, M. (1990). Born's probabilistic interpretation: a case study of 'concepts in flux'. *Studies in the History and Philosophy of Science*, **21**, 563–88.

Beller, M. (1992a). The birth of Bohr's complementarity: the context and the dialogues. *Studies in the History and Philosophy of Science*, **23**, 147–80.

Beller, M. (1992b). The genesis of Bohr's complementarity principle and the Bohr–Heisenberg dialogue. In *The Scientific Enterprise. The Bar-Hillel Colloquium: Studies in the History, Philosophy, and Sociology of Science*, vol. 4, ed. E. Ullman-Margalit. Dordrecht: Kluwer Academic Publishers, pp. 273–95.

Beller, M. (1999). *Quantum Dialogue: The Making of a Revolution.* Chicago: University of Chicago.

Bitbol, M. (1996). *Schrödinger's Philosophy of Quantum Mechanics.* Dordrecht: Kluwer Academic Publishers.

Bitbol, M. and Darrigol, O. eds. (1992). *Erwin Schrödinger: Philosophy and the Birth of Quantum Mechanics.* Paris: Éditions Frontièrs.

Blokhintzev, D.I. (1952). Critique de la conception idéaliste de la théorie quantique. In *Questions Scientifiques*, vol. 1: Physique, trans. F. Lurçat. Paris: Les éditions de la nouvelle critique, pp. 95–129.

Bohr, N. (1928). The quantum postulate and the recent development of atomic theory. *Nature*, **121**, 580–90.

Bohr, N. (1931). Maxwell and modern theoretical physics. *Nature*, **128**, 691–2.

Bohr, N. (1932). Chemistry and the quantum theory of atomic constitution. *Journal of Chemical Society*, 349–84.

Bohr, N. (1935). Can quantum-mechanical description of physical reality be considered complete? *Physical Review*, **48**, 696–702.

Bohr, N. (1937). Causality and complementarity. *Philosophy of Science*, **4**, 289–98.

Bohr, N. (1948). On the notions of causality and complementarity. *Dialectica*, **1**, 312–19.

Bohr, N. (1949). Discussion with Einstein on epistemological problems in atomic physics. In *Albert Einstein: Philosopher-Scientist*, vol. 7: The Library of Living Philosophers, ed. P.A. Schilpp. La Salle: Open Court, pp. 201–41.

Bohr, N. (1984). *Collected Works*, vol. 5: The Emergence of Quantum Mechanics 1924–1926, ed. K. Stolzenburg. Amsterdam: North-Holland.

Bohr, N. (1985). *Collected Works*, vol. 6: Foundations of Quantum Mechanics I 1926–1932, ed. J. Kalckar. Amsterdam: North-Holland.

Bohr, N. (1987a). Introductory survey. In *Atomic Theory and the Description of Nature. The Philosophical Writings of Niels Bohr*, vol. 1. Woodbridge, CT: Ox Box Press, pp. 1–24.

Bohr, N. (1987b). The quantum of action and the description of nature. In *Atomic Theory and the Description of Nature. The Philosophical Writings of Niels Bohr*, vol. 1. Woodbridge, CT: Ox Bow Press, pp. 92–101.

Bohr, N. (1996a). The causality problem in atomic physics [1938]. In *Collected Works*, vol. 7: Foundations of Quantum Mechanics II 1933–1958, ed. J. Kalckar. Amsterdam: Elsevier, pp. 303–22.

Bohr, N. (1996b). *Collected Works*, vol. 7: Foundations of Quantum Mechanics II 1933–1958, ed. J. Kalckar. Amsterdam: Elsevier.

Bohr, N., Kramers, H.A. and Slater, J.C. (1924). The quantum theory of radiation. *Philosophical Magazine and Journal of Science*, **47**, 785–802.

Bokulich, A. (2004). Open or closed? Dirac, Heisenberg, and the relation between classical and quantum mechanics. *Studies in History and Philosophy of Modern Physics*, **35**, 377–96.

Bokulich, A. (2006). Heisenberg meets Kuhn: closed theories and paradigms. *Philosophy of Science*, **78**, 90–107.

Born, M. (1923). Quantentheorie und Störungsrechnung. *Naturwissenschaften*, **11**, 537–42.

Born, M. (1925). *Vorlesungen über Atommechanik*. Berlin: J. Springer-Verlag. Trans. as *Mechanics of the Atom*, J.W. Fisher. London: Bell.

Born, M. (1926a). Quantenmechanik der Stossvorgänge. *Zeitschrift für Physik*, **38**, 803–27. Trans. as M. Born, 1968b.

Born, M. (1926b). Zur Quantenmechanik der Stossvorgänge. Zeitschrift für Physik, **37**, 863–7. Trans. as M. Born, 1983.

Born, M. (1927). Physical aspects of quantum mechanics. *Nature*, **119**, 354–7.

Born, M. (1956a). The interpretation of quantum mechanics. In *Physics in My Generation. A Selection of Papers*. London: Pergamon Press, pp. 140–50.

Born, M. (1956b). Physics and metaphysics. In *Physics in My Generation. A Selection of Papers*. London: Pergamon Press, pp. 93–108.

Born, M. (1956c). Statistical interpretation of quantum mechanics. In *Physics in my Generation. A Selection of Papers*. London: Pergamon Press, pp. 177–88.

Born, M. (1960). *Problems of Atomic Dynamics*. New York: Friedrick Ungar.

Born, M. (1968). Quantum mechanics of collision processes. In *Wave Mechanics*, ed. G. Ludwig. Oxford: Pergamon Press, pp. 206–25.

Born, M. (1983). On the quantum mechanics of collisions. In *Quantum Theory and Measurement*, ed. J.A. Wheeler and W.H. Zurek. Princeton: Princeton University Press, pp. 52–5.

Born, M. and Heisenberg, W. (1923). Die Elektronenbahn in angeregten Heliumatom. *Zeitschrift für Physik*, **16**, 229–43.

Born, M., Heisenberg, W. and Jordan, P. (1926). Zur Quantenmechanik II. *Zeitschrift für Physik*, **35**, 557–615. Trans. as M. Bohr, W. Heisenberg and P. Jordan, 1967.

Born, M., Heisenberg, W. and Jordan, P. (1967). On quantum mechanics II. In *Sources of Quantum Mechanics*, ed. B.L. van der Waerden. Amsterdam: North-Holland, pp. 321–85.

Born, M. and Jordan, P. (1925). Zur Quantentheorie aperiodische Vorgänge. *Zeitschrift für Physik*, **33**, 479–505.

Bridgman, P. (1927). *The Logic of Modern Physics*. New York: MacMillan.

Bridgman, P. (1936). *The Nature of Physical Theory*. New York: Dover.

Bridgman, P. (1949). Einstein's theories and the operational point of view. In *Albert Einstein: Philosopher-Scientist*, ed. P. Schilpp. La Salle: Open Court, pp. 333–54.

de Broglie, L. (1928a). The principles of the new wave mechanics. In *Selected Papers on Wave Mechanics*, ed. L. de Broglie and L. Brillouin, trans. W.M. Deans. London: Blackie & Son Ltd, pp. 55–78.

de Broglie, L. (1928b). The wave mechanics and the atomic structure of matter and of radiation [1927]. In *Selected Papers on Wave Mechanics*, ed. L. de Broglie and L. Brillouin, trans. W.M. Deans. London: Blackie & Son Ltd, pp. 113–38.

Bromberg, J. (1977). Dirac's quantum electrodynamics and the wave–particle equivalence. In *History of Twentieth Century Physics*, ed. C. Weiner. New York: Academic Press, pp. 147–57.

Bub, J. (1995). Complementarity and the orthodox (Dirac-von Neumann) interpretation of quantum mechanics. In *Perspectives on Quantum Reality: Non-Relativistic, Relativistic, and Field-Theoretic*, ed. R. Clifton. Dordrecht: Kluwer, pp. 211–26.

Buschhorn, G.W. and Wess, J. eds. (2004). *Fundamental Physics – Heisenberg and Beyond* – Proceedings from the Werner Heisenberg Centennial Symposium 'Developments in Modern Physics' (2001). Berlin; New York: Springer.

Camilleri, K. (2005). Heisenberg and the transformation of Kantian philosophy. *International Studies in the Philosophy of Science*, **19**(3), 271–87.

Camilleri, K. (2006). Heisenberg and the wave–particle duality. *Studies in the History and Philosophy of Modern Physics*, **37**, 298–315.

Camilleri, K. (2007a). Bohr, Heisenberg and the divergent viewpoints of complementarity. *Studies in the History and Philosophy of Modern Physics*, **38**, 514–28.

Camilleri, K. (2007b). Indeterminacy and the limits of classical concepts: the turning point in Heisenberg's thought. *Perspectives on Science* **15**(2), 176–99.

Carazza, B. and Kragh, H. (1995). Heisenberg's lattice world: the 1930 sketch. *American Journal of Physics*, **63**, 595–605.

Carella, M.J. (1976). Heisenberg's concept of matter as potency. *Diogenes* (winter edn), **96**, 25–37.

Carnap, R. (1922). *Der Raum: eine Beitrag zur Wissenschaftslehre*. Berlin: Reuther and Reichard.

Carnap, R. (1975). Observation language and theoretical language. In *Rudolf Carnap, Logical Empiricist*, ed. J. Hintikka. Dordrecht: D. Reidel, pp. 77–85.

182   *References*

Carson, C.L. (1995). *Particle Physics and Cultural Politics: Werner Heisenberg and the Shaping of a Role for the Physicist in Post-war Germany.* Unpublished PhD thesis, Harvard University.

Cartwright, N. (1987). Max Born and the reality of quantum probabilities. In *The Probabilistic Revolution*, vol. 2: Ideas in the Sciences, ed. L. Kruge, G. Gigerenzer and H.S. Morgan. Cambridge: MIT Press, pp. 409–16.

Casimir, H.B.G. (1986). Epistemological considerations. In *The Lesson of Quantum Theory.* Niels Bohr Centenary Symposium, ed. J. de Boer, E. Dal and O. Ulfbeck. New York: Elsevier Science Publishers, pp. 13–20.

Cassidy, D.C. (1976). *Werner Heisenberg and the Crisis of Quantum Theory 1920–1924.* Unpublished PhD thesis, Purdue University.

Cassidy, D.C. (1979). Heisenberg's first core model of the atom: the formation of a professional style. *Historical Studies in the Physical Sciences*, **10**, 187–24.

Cassidy, D.C. (1992). *Uncertainty: The Life and Science of Werner Heisenberg.* New York: W.H. Freeman.

Cassidy, D.C. (1998). Answer to the question: when did the indeterminacy principle become the uncertainty principle? *American Journal of Physics*, **66**, 278–9.

Cassirer, E. (1923). *Substance and Function and Einstein's Theory of Relativity*, trans. W.C. Swabey and M.C. Swabey. Chicago: Open Court.

Cassirer, E. (1955). *The Philosophy of Symbolic Forms*, vol. 1: Language, trans. R. Manheim. New Haven: Yale University Press.

Cassirer, E. (1956). *Indeterminism and Determinism in Modern Physics*, trans. O. Theodore Benfey. New Haven: Yale University Press.

Chevalley, C. (1988). Physical reality and closed theories in Werner Heisenberg's early papers. In *Theory and Experiment*, ed. D. Batens and J.P. van Bendegem. Dordrecht: D. Reidel, pp. 159–76.

Chevalley, C. (1991a). Glossaire. In *Physique Atomique et Conaissance Humaine par N. Bohr*, Trad. E. Bauer and R. Omnes, revue par C. Chevalley. Paris: Guillimard, pp. 345–567.

Chevalley, C. (1991b). La Signification Physique de la Théorie Quantique en 1926: Heisenberg et Schrödinger. *Revue du Palais de la Découverte*, **40**, 49–61.

Chevalley, C. (1992). Physique Quantique et Philosophie. *Le Débat*, **72**, 65–76.

Chevalley, C. (1994). Niels Bohr's words and the Atlantis of Kantianism. In *Niels Bohr and Contemporary Philosophy*, ed. J. Faye and H. Folse. Dordrecht: Kluwer Academic Publishers, pp. 33–55.

Chevalley, C. (1998). Introduction. In *Philosophie: Le Manuscrit de 1942* by W. Heisenberg, trans. C. Chevalley. Paris: Éditions du Seuil, pp. 21–245.

Chevalley, C. (1999). Why do we find Bohr obscure? In *Epistemological and experimental perspectives on quantum physics*, ed. D. Greenberger, W.L. Reiter and A. Zeilinger. Dordrecht; London: Kluwer Academic, pp. 59–73.

Coffa, J.A. (1991). *The Semantic Tradition from Kant to Carnap: To the Vienna Station*, ed. L. Wessels. Cambridge: Cambridge University Press.

Cushing, J. (1994). *Quantum Mechanics: Historical Contingency and Copenhagen Hegemony.* Chicago: University of Chicago Press.

Darrigol, O. (1986). The origin of quantized matter waves. *Historical Studies in the Physical and Biological Sciences*, **16**, part 2, 197–253.

Darrigol, O. (1992). *From C-Numbers to Q-Numbers: Classical Analogy in the History of Quantum Theory*. Berkeley: University of California Press.

Darwin, C.G. (1927). Free motion in the wave mechanics. *Proceedings of the Royal Society*, **A117**, 258–93.

Degen, A. (1989). *Interpretations of Quantum Physics, the Mystical, and the Paranormal: Einstein, Schrödinger, Bohr, Pauli and Jordan*. Unpublished PhD thesis, Drew University.

Dirac, P. (1927a). The physical interpretation of quantum dynamics. *Proceedingas of the Royal Society*, **A113**, 621–41.

Dirac, P. (1927b). The quantum theory of emission and absorption of radiation. *Proceedings of the Royal Society*, **A114**, 243–65.

Dirac, P. (1977). Recollections of an exciting era. *History of Twentieth Century Physics*, ed. C. Weiner. New York: Academic Press, pp. 109–48.

Diskussion über Kausalität und Quantenmechanik. (1931). *Erkenntnis*, **2**, 183–8.

Ehrenfest, P. (1927). Bemerkung über die angenäherte Gültigkeit der klassischen Mechanik innerhalb der Quantenmechanik. *Zeitschrift fur Physik*, **45**, 455–7.

Einstein, A. (1916). Ernst Mach. *Physicalische Zeitschrift*, **17**, 101–4.

Einstein, A. (1920). *Relativity, The Special and the General Theory: A Popular Exposition*, 5th edn, trans. R.W. Lawson. London: Methuen.

Einstein, A. (1971). *The Born–Einstein Letters: Correspondence Between Albert Einstein and Max and Hedwig Born from 1916 to 1955*, trans. I. Born. New York: McMillan.

Einstein, A. (1996). Geometry and experience. In *Logical Empiricism at its Peak: Schlick, Carnap and Neurath*, vol. 2: Science and Philosophy in the Twentieth Century, Basic Works of Logical Empiricism, ed. S. Sarkar. New York: Garland, pp. 97–126.

Einstein, A., Podolosky, B. and Rosen, N. (1935). Can a quantum-mechanical description of physical reality be considered complete? *Physical Review*, **47**, 777–80.

Électrons et Photons. (1928). *Rapports et Discussions du Cinquieme Conseil de Physique, tenu a Buxelles du 24 au 29 Octobre 1927 sous les auspices de l'Institut Internationale de Physique Solvay*. Paris: Gauthiers-Villars.

Favrholdt, D. (1992). *Niels Bohr's Philosophical Background*. Copenhagen: Munksgaard.

Favrholdt, D. (1994). Niels Bohr and realism. In *Niels Bohr and Contemporary Philosophy*, ed. J. Faye and H. Folse. Dordrecht: Kluwer Academic Publishers, pp. 77–96.

Faye, J. (1991). *Niels Bohr: His Heritage and Legacy: An Anti-Realist View of Quantum Mechanics*. Dordrecht: Kluwer Academic Publishers.

Faye, J. and Folse, H.J. eds. (1994). *Niels Bohr and Contemporary Philosophy*. Dordrecht: Kluwer Academic Publishers.

Fine, A. (1986). *The Shaky Game: Einstein, Realism and the Quantum Theory*. Chicago: University of Chicago Press.

Fock, V.I. (1957). On the interpretation of quantum mechanics. *Czechoslovak Journal of Physics*, **7**, 643–56.

Folse, H.J. (1978). Kantian aspects of complementarity. *Kant Studien*, **69**, 58–66.

Folse, H.J. (1985). *The Philosophy of Niels Bohr: The Framework of Complementarity*. Amsterdam: North-Holland, Elsevier Science.

Folse, H.J. (1987). Niels Bohr's concept of reality. In *Symposium on the Foundations of Modern Physics. The Copenhagen Interpretation 60 years after the Como Lecture*, ed. P. Lahti and P. Mittelstaedt. Singapore: World Scientific, pp. 161–79.

Forman, P. (1984). *Kausalität, Anshaulichkeit* and *Individualität*, or How cultural values prescribe the character and lessons ascribed to quantum mechanics. In *Society and Knowledge: Contemporary Perspectives in Sociology of Knowledge*, ed. N. Stehr and V. Meja. New Brunswick: Transaction Books, pp. 333–47.

Frank, P. (1928). Über die 'Anschaulichkeit' physikalische Theorien. *Naturwissenschaften*, **8**, 121–6.

Frank, P. (1957). *Philosophy of Science: The Link Between Science and Philosophy*. Englewood Cliffs, NJ: Prentice Hall.

Frank, P. (1975). *Modern Science and its Philosophy*. New York: Adorno Press.

Frappier, M. (2004). *Heisenberg's Notion of Interpretation*. Unpublished PhD thesis, University of Western Ontario.

Friedman, M. (1994). Geometry, convention, and the relativized a priori: Reichenbach, Schlick and Carnap. In *Logic, Language, and the Structure of Scientific Theories*, ed. W. Salmon and G. Wolters. Pittsburgh: University of Pittsburgh Press, pp. 21–34.

Friedman, M. (1996). Epistemology in the *Aufbau*. In *The Emergence of Logical Empiricism: From 1900 to the Vienna Circle*, vol. 1: Science and Philosophy in the Twentieth Century, Basic Works of Logical Empiricism, ed. S. Sarkar. New York: Garland, pp. 367–409.

Friedman, M. (1999). *Reconsidering Logical Positivism*. Cambridge: Cambridge University Press.

Gadamer, H.G. (1989). *Truth and Method*. New York: Crossroad.

Gembillo, G. (1987). *Werner Heisenberg: la filosofia di un fisico*. Napoli: Gianni.

Gembillo, G. and Altavilla, C. eds. (2002). *Werner Heisenberg: Scienzato e filosofo*. Messina: Armando Siciliano.

Geyer, B., Herwig, H. and Rechenberg, H. eds. (1993). *Werner Heisenberg: Physiker und Philosoph*. Leipzig: Spektrum.

Gieser, S. (2005). *Innermost Kernel: Depth Psychology and Quantum Physics*. Wolfgang Pauli's Dialogue with C.G. Jung. Berlin: Springer.

Götsch, J. ed. (1992). *Erwin Schrödinger's World View: The Dynamics of Knowledge and Reality*. Dordrecht: Kluwer Academic Publishers.

Gower, B. (2000). Cassirer, Schlick and 'structural' realism: the philosophy of the exact sciences in the background to early logical empiricism. *British Journal for the History of Philosophy*, **8**(1), 71–106.

Habermas, J. (1999). Hermeneutic and analytic philosophy. Two complementary versions of the linguistic turn. In *German Philosophy Since Kant*, ed. A. O'Hear. Cambridge: Cambridge University Press, pp. 413–41.

Heelan, P. (1965). *Quantum Mechanics and Objectivity: A Study of the Physical Philosophy of Werner Heisenberg*. The Hague: Martinus Nijhoff.

Heelan, P. (1975). Heisenberg and radical theoretical change. *Zeitschrift für Allgemeine Wissenschaftstheorie*, **6**, 113–38.

Heidegger, M. (1971). *On the Way to Language*, Trans. P.D. Herz. New York: Harper and Row.

Heidegger, M. (1977). *Dem Andenken Martin Heideggers: zum 26 Mai 1976*, Frankfurt: Am Main.

Heisenberg, E. (1984). *Inner Exile. Recollections of a Life with Werner Heisenberg*, trans. S. Cappellari and C. Morris. Boston: Bikhäuser.

Heisenberg, W. (1925a). Über quantentheoretische Umdeutung kinematischer und mechanischer Bezeihungen. *Zeitschrift für Physik*, **33**, 879–93. Trans. as W. Heisenberg, 1967b.

Heisenberg, W. (1925b). Zur Quantentheorie der Linienstruktur und der anomalen Zeemaneffekte. *Zeitschrift für Physik*, **32**, 841–60.

Heisenberg, W. (1926a). Mehrkörperproblem und Resonanz in der Quantenmechanik, *Zeitschrift für Physik*, **38**, 411–26.

Heisenberg, W. (1926b). Quantenmechanik. *Naturwissenschaften*, **14**, 989–94.

Heisenberg, W. (1926c). Schwankungserscheinungen und Quantenmechanik. *Zeitschrift für Physik*, **40**, 501–6.

Heisenberg, W. (1926d). Über quantentheoretische Kinematik und Mechanik. *Mathematische Annalen*, **95**, 683–705.

Heisenberg, W. (1927a). Über den anschaulichen Inhalt der quantentheoretische Kinematik und Mechanik. *Zeitschrift für Physik*, **43**, 172–98. Trans. as W. Heisenberg, 1983.

Heisenberg, W. (1927b). Über die Grundprinzipien der 'Quantenmechanik'. *Forschungen und Fortschriftte*, **3**, 83.

Heisenberg, W. (1929). Die Entwicklung der Quantentheorie 1918–1928. *Die Naturwissenschaften*, **17**, 490–6.

Heisenberg, W. (1930). *The Physical Principles of the Quantum Theory*, trans. C. Eckhart and F.C. Hoyt. Chicago: University of Chicago Press.

Heisenberg, W. (1931a). Die Kausalgesetz und Quantenmechanik. *Erkenntnis*, **2**, 172–82.

Heisenberg, W. (1931b). Die Rolle der Unbestimmtheitsrelationen in der modernen Physik. *Monatshefte für Mathematik und Physik*, **38**, 365–72.

Heisenberg, W. (1934). Wandlungen der Grundlagen der exakten Naturwissenschaft in jungster Zeit. *Angewandte Chemie*, **47**, 697–702. Trans. as Heisenberg, 1952c.

Heisenberg, W. (1936). Prinzipielle Fragen der modernen Physik. In *Neue Fortschritte in der exakten Wissenschaften. Fünf Weiner Vorträge*. Leipzig: Franz Deuticke, pp. 91–102. Trans. as Heisenberg, 1952a.

Heisenberg, W. (1948). Der Begriff 'abgeschlossene Theorie' in der modernen Naturwissenschaft. *Dialectica*, **2**, 331–6.

Heisenberg, W. (1949). *Wandlungen in den Grundlagen der Naturwissenschaft*, 4th edn. Leipzig: S. Hirzel Verlag.

Heisenberg, W. (1952a). Questions of principle in modern physics. In *Philosophic Problems in Nuclear Science*, trans. F.C. Hayes. London: Faber and Faber, pp. 41–52.

Heisenberg, W. (1952b). On the history of the physical interpretation of nature. In *Philosophic Problems of Nuclear Science*, trans. F.C. Hayes. London: Faber and Faber, pp. 27–40.

Heisenberg, W. (1952c). Recent changes in the foundations of exact science. In *Philosophic Problems in Nuclear Science*, trans. F.C. Hayes. London: Faber and Faber, pp. 11–26.

Heisenberg, W. (1955). The development of the interpretation of quantum theory. In *Niels Bohr and the Development of Physics*. Essays Dedicated to Niels Bohr on the

Occasion of his Seventieth Birthday, ed. W. Pauli, L. Rosenfeld and V. Weisskopf. New York: McGraw Hill, pp. 12–29.

Heisenberg, W. (1958a). Atomic physics and causal law. In *The Physicist's Conception of Nature*, trans. A.J. Pomerans. London: Hutchinson Scientific and Technical, pp. 32–50.

Heisenberg, W. (1958b). The idea of nature in contemporary physics. In *The Physicist's Conception of Nature*, trans. A.J. Pomerans. London: Hutchinson Scientific and Technical, pp. 7–31.

Heisenberg, W. (1958c). *The Physicist's Conception of Nature*, trans. A.J. Pomerans. London: Hutchinson Scientific and Technical.

Heisenberg, W. (1958d). *Physics and Philosophy: The Revolution in Modern Science*. Gifford Lectures in the winter of 1955–6. London: George Allen and Unwin.

Heisenberg, W. (1959). Grundlegende Voraussetzungen in der Physik der Elementarteilchen. In *Martin Heidegger zum Siebzigsten Geburtstag Festschrift*, ed. G. Neske. Pfullingen: Verlag, pp. 291–7.

Heisenberg, W. (1960a). Erinnerungen an die Zeit der Entwicklung der Quantenmechanik. In *Theoretical Physics in the Twentieth Century*, A Memorial Volume to Wolfgang Pauli, ed. M. Fierz and V.F. Weisskopf. New York: Interscience Publishers, pp. 40–7.

Heisenberg, W. (1960b). Sprache und Wirklichkeit in der modernen Physik. In *Gestalt und Gedanke Band 6, Jahrbuch der Bayerischen Akademie der schönen Künste*, ed. Bayerischen Akademie der schönen Künste. Munich: R. Oldenbourg, pp. 32–62.

Heisenberg, W. (1965). The development of quantum mechanics. In *Nobel Lectures, Including Presentation Speeches and Laureates' Biographies. Physics 1922–1941*, ed. The Nobel Foundation. Amsterdam: Elsevier, pp. 290–301.

Heisenberg, W. (1967a). Quantum-theoretical re-interpretation of kinematic and mechanical relations. In *Sources of Quantum Mechanics*, ed. B.L. van der Waerden. Amsterdam: North-Holland, pp. 261–76.

Heisenberg, W. (1967b). Quantum theory and its interpretation. In *Niels Bohr: His Life and Work as Seen by His Friends*, ed. S. Rozental. Amsterdam: North-Holland, pp. 94–108.

Heisenberg, W. (1968). Theory, criticism and a philosophy. In *From a Life of Physics*. Evening Lectures at the International Centre for Theoretical Physics, ed. H.A. Bethe. Trieste, Italy, Wein: IAEA, pp. 31–46.

Heisenberg, W. (1971). *Physics and Beyond*. New York: Harper & Row.

Heisenberg, W. (1977). Remarks about the uncertainty principle. In *The Uncertainty Principle and the Foundations of Quantum Mechanics*, ed. W.C. Price and S. Chissick. New York: Wiley, pp. 3–6.

Heisenberg, W. (1983). The physical content of quantum kinematics and mechanics. In *Quantum Theory and Measurement*, ed. J.A. Wheeler and W.H. Zurek. Princeton: Princeton University Press, pp. 62–84.

Heisenberg, W. (1984a). Erkenntnistheoretische Probleme in der modernen Physik. In *Gesammelte Werke*, Series C: Philosophical and Popular Writings, vol. I: Physik und Erkenntnis 1927–1955, ed. W. Blum, H. Dürr and H. Rechenberg. Munich: Piper, pp. 22–8.

Heisenberg, W. (1984b). *Gesammelte Werke*, Series B: Scientific Review Papers, Talks and Books, ed. W. Blum, H. Dürr and H. Rechenberg. Berlin: Springer-Verlag.

Heisenberg, W. (1984c). *Gesammelte Werke*, Series C: Philosophical and Popular Writings, vol. I: Physik und Erkenntnis 1927–1955, ed. W. Blum, H. Dürr and H. Rechenberg. Munich: Piper.

Heisenberg, W. (1984d). *Gesammelte Werke*, Series C: Philosophical and Popular Writings, Vol. II: Physik und Erkenntnis 1956–1968, ed. W. Blum, H.P. Dürr and H. Rechenberg. Zurich: Piper.

Heisenberg, W. (1984e). Ordnung der Wirklichkeit. In *Gesammelte Werke*, Series C: Philosophical and Popular Writings, Vol. I: Physik und Erkenntnis 1927–1955, ed. W. Blum, H. Dürr and Helmet Rechenberg. Munich: Piper, pp. 217–306.

Heisenberg, W. (1985a). *Gesammelte Werke*, Series C: Philosophical and Popular Writings, Vol. III: Physik und Erkenntnis 1969–1976, ed. W. Blum, H. Dürr and H. Rechenberg. Munich: Piper.

Heisenberg, W. (1985b). Is ene deterministische Ergänzung der Quantenthmechanik möglich? [1935]. In *Wolfgang Pauli. Wissenschaftlicher Briefwechsel mit Bohr, Einstein, Heisenberg*, vol. 1: 1919–1929, ed. A. Hermann, K. von Meyenn and V.F. Weisskopf. New York: Springer-Verlag, pp. 409–18.

Heisenberg, W. (1989). Development of concepts in the history of quantum mechanics. In *Encounters with Einstein: And Other People, Places and Particles*. Princeton: Princeton University Press, pp. 19–36.

Heisenberg, W. (1991). *Indeterminazione e Realtà*, ed. and trans. G. Gembillo. Naples: Guida Editori.

Heisenberg, W. (1998). *Philosophie: Le Manuscrit de 1942*, Introduction et traduction de Catherine Chevalley. Paris: Éditions du Seuil.

Heitler, W. (1949). The Departure from Classical Thought in Modern Physics. In *Albert Einstein: Philosopher-Scientist*, ed. P.A. Schilpp. Evanston; Illinois: George Banta Publishing Company, pp. 179–98.

Held, C. (1994). The meaning of complementarity. *Studies in History and Philosophy of Science*, **25**, 871–93.

Helmholtz, H. von. (1876). The origin and meaning of geometrical axioms. *Mind*, **1**, 301–21.

Helmholtz, H. von. (1878). The origin and meaning of geometrical axioms (II). *Mind*, **3**, 212–224. Trans. into German as Helmholtz, 1883.

Helmholtz, H. von. (1883). Über den Urpsrung und Sinn der geometrische Sätze; Antwort gegen Hernn Professor Land. In *Wissenschaftliche Abhandlungen*, vol. 2. Leipzig: Johann Ambrosius Barth, pp. 640–60.

Helmholtz, H. von. (1977). *Epistemological Writings*, trans. M.F. Lowe, R.S. Cohen and Y. Elkana, ed. M. Schlick and P. Herz. Dordrecht: D. Reidel.

Hempel, H.-P. (1990). *Natur und Geschichte: Die Jahrhudertdialog zwischen Heidegger und Heisenberg*. Frankfurt Am Main: A. Hain.

Hendry, J. (1984). *The Creation of Quantum Mechanics and the Bohr–Pauli Dialogue*. Dordrecht: D. Reidel.

Hentschel, K. (1998). Heinrich Hertz' mechanics: a model for Werner Heisenberg's April 1925 paper on the anomalous Zeeman effect. In *Heinrich Hertz: Classical Physicist: Modern Philosopher*, ed. D. Bairol, R.I.G. Hughes and A. Nordman. Dordrecht: Kluwer Academic Publishers, pp. 183–223.

Hermann, A. (1977). *Die Jahrhundertwissenschaft: Werner Heisenberg und die Physik Seiner Zeit*. Stuttgart: Deutsche Verlags-Anstalt.

Hermann, G. (1935a). *Die Naturphilosophischen Grundlagen der Quantenmechanik.* Berlin: Öffentliches Leben.

Hermann, G. (1935b). Die naturphilosophischen Grundlagen der Quantenmechanik. *Die Naturwissenschaften*, **23**(42), 718–21.

Hermann, G. (1937). Über die grundlagen physikalischer Aussagen in den älteren und der modernen Theorien. *Abhandlungen der Fries'schen Schule, Neue Folge*, **6**(3/4), 309–98.

Hermann, G., May, E. and Vogel, T. (1937). *Die Bedeutung der modern Physik für die Theorie der Erkenntnis.* Hirzel: Leipzig.

Hilgevoord, J. and Uffink, J. (1988). The mathematical expression of the uncertainty principle. In *Microphysical Reality and Quantum Description*, ed. A. van der Merwe, G. Tarozzi and F. Selleri. Dordrecht: Kluwer, pp. 91–114.

Hilgevoord, J. and Uffink, J. (1990). A new view on the uncertainty principle. In *Sixty-Two Years of Uncertainty, Historical and Physical Inquiries into the Foundations of Quantum Mechanics*, ed. A.E. Miller. New York: Plenum, pp. 121–39.

Holland, P. (1993). *The Quantum Theory of Motion: An Account of the de Broglie-Bohm Causal Interpretation of Quantum Mechanics.* Cambridge: Cambridge University Press.

Honner, J. (1982). The transcendental philosophy of Niels Bohr. *Studies in the History and Philosophy of Science*, **13**, 1–29.

Honner, J. (1987). *The Description of Nature: Niels Bohr and the Philosophy of Quantum Physics.* Oxford: Clarendon Press, Oxford University Press.

Hooker, C. (1972). The nature of quantum mechanical reality: Einstein vs Bohr. In *Paradigms and Paradoxes: The Philosophical Challenge of the Quantum Domain*. vol. 5: Pittsburgh Studies in the Philosophy of Science. Pittsburgh: University of Pittsburgh Press, pp. 67–302.

Howard, D. (1985). Einstein on locality and separability. *Studies in the History and Philosophy of Science*, **16**, 171–201.

Howard, D. (1994a). Einstein, Kant, and the origins of logical empiricism. In *Logic, Language, and the Structure of Scientific Theories*, ed. W. Salmon and G. Wolters. Pittsburgh: University of Pittsburgh Press, pp. 45–105.

Howard, D. (1994b). What makes a classical concept classical? Toward a reconstruction of Niels Bohr's philosophy of physics. In *Niels Bohr and Contemporary Philosophy*, ed. J. Faye and H. Folse. Dordrecht: Kluwer Academic Publishers, pp. 201–29.

Howard, D. (2004). Who invented the 'Copenhagen interpretation'? A study in mythology. *Philosophy of Science*, **71**, 669–82.

Howard, D. (in press). Physics and the philosophy of science at the turn of the twentieth century. Forthcoming in the *Enciclopedia Italiana di Storia della Scienza* under the title, 'Fisica e Filosofia della Scienza all'Alba del XX Secolo'.

Hörz, H. (1968). *Werner Heisenberg und die Philosophie.* Berlin: VEB Deutscher Verlag der Wissenschaften.

Hund, F. (1974). *The History of Quantum Theory*, trans. G. Reece. London: Harrap.

Jammer, M. (1966). *The Conceptual Development of Quantum Mechanics.* New York: McGraw-Hill.

Jammer, M. (1974). *The Philosophy of Quantum Mechanics: The Interpretations of Quantum Mechanics in Historical Perspective.* New York: Wiley.

Jordan, P. (1927a). Philosophical foundations of quantum theory. *Nature*, **119**, 556–8.

Jordan, P. (1927b). Über eine neue Begründung der Quantenmechanik. *Zeitschrift für Physik*, **39**, 809–38.

Jordan, P. (1927c). Zur Quantenmechanik der Gasentartung. *Zeitschrift für Physik*, **44**, 473–80.

Jordan, P. (1936). *Anschauliche Quantentheorie. Eine Einführung in die moderne Auffassung der Quantenerscheinungen.* Berlin: Verlag von Julius Springer.

Jordan, P. (1944). *Physics of the Twentieth Century.* New York: Philosophical Library.

Jordan, P. and Klein, O. (1927). Zum Mehrkörperproblem der Quantentheorie. *Zeitschrift für Physik*, **45**, 751–65.

Jordan, P. and Wigner, E. (1928). Über das Paulische Aquivalenzverbot. *Zeitschrift für Physik*, **47**, 631–51.

Kaiser, D. (1992). More roots of complementarity: Kantian aspects and influences. *Studies in the History and Philosophy of Science*, **23**, 213–39.

Kant, I. (1934). *Immanuel Kant's Critique of Pure Reason*, trans. N.K. Smith, abridged edn. London: MacMillan and Co.

Kennard, E.H. (1927). Zur Quantenmechanik einfacher Bewegungstypen. *Zeitschrift für Physik*, **44**, 326–52.

Kennard, E.H. (1928). On the quantum mechanics of a system of particles. *Physical Review*, **31**, 876–90.

Kleint, C. and Wiemers, G. eds. (1993). *Werner Heisenberg in Leipzig, 1927–1942.* Berlin: Akademie Verlag.

Kojevnikov, A. (2002). The last century of physics, essay review. *Annals of Science*, **59**, 419–22.

Konno, H. (1978). The historical roots of Born's probabilistic interpretation. *Japaenese Studies in the History of Science*, **17**, 129–45.

Kramers, H.A. and Heisenberg, W. (1925). Über die Steuung von Strahlung durch Atome. *Zeitschrift für Physik*, **31**, 681–708.

Kronig, R. (1960). The turning point. In *Theoretical Physics in the Twentieth Century: A Memorial Volume to Wolfgang Pauli*, ed. M. Fierz and V.F. Weisskopf. New York: Interscience Publishers, pp. 5–39.

Lacki, J. (2002). Observability, *Anschaulichkeit* and abstraction: a journey into Werner Heisenberg's Science and Philosophy. *Fortschriften Physik*, **50**, 440–58.

Ladyman, J. (1998). What is structural realism? *Studies in the History and Philosophy of Science*, **29**, 409–24.

Lafont, C. (1994). World disclosure and reference. *Thesis Eleven*, **37**, 46–63.

Lafont, C. (1999). *The Linguistic Turn in Hermeneutic Philosophy*, trans. J. Medina. Cambridge: MIT Press.

Landé, A. (1926). Neue Wege in der Quantentheorie. *Naturwissenschaften*, **14**, 455–8.

Laue, M. von. (1932). Zu den Erörterungen über Kausalität. *Naturwissenschaften*, **20**, 915–16.

Laue, M. von. (1934). Über Heisenbergs Ungenauigzeitbeziehungen und ihre erkenntnistheoretische Bedeutung. *Naturwissenschaften*, **26**, 439–41.

Lewis, C.I. (1929). *Mind and World Order: Outline of a Theory of Knowledge.* New York: Dover.

Liesenfeld, C. (1992). *Philosophische Weltbilder des 20. Jahrhunderts: Eine Interdisziplinäre Studie zu Max Planck und Werner Heisenberg.* Würzburg: Königshausen and Neumann.

London, F. and Bauer, E. (1983). The theory of observation in quantum mechanics. In *Quantum Theory and Measurement*, ed. J.A. Wheeler and W.H. Zurek. Princeton: Princeton University Press, pp. 217–59.

Lorenz, K. (1977). Kant's doctrine of the a priori in the light of modern biology. In *Behind the Mirror: A Search for a Natural History of Human Knowledge*, trans. R. Taylor. New York: Harcourt Brace Jovanovich, pp. 181–217.

MacKinnon, E. (1977). Heisenberg, models and the rise of matrix mechanics. *Historical Studies in the Physical Sciences*, **8**, 137–88.

MacKinnon, E. (1982). *Scientific Explanation and Atomic Physics*. Chicago: The University of Chicago Press.

MacKinnon, E. (1984). Semantics and quantum logic. In *Science and Reality: Recent Work in the Philosophy of Science, Essays in Honor of Ernan McMullin*, ed. J.T. Cushing, C.F. Delaney and G.M. Gutting. Notre Dame: University of Notre Dame Press, pp. 173–95.

MacKinnon, E. (1985). Bohr on the foundations of quantum theory. In *Niels Bohr: A Centenary Volume*, ed. A.P. French and P.J. Kennedy. Cambridge: Harvard University Press, pp. 101–120.

McMullin, E. (1954). *The Principle of Uncertainty: A Preliminary Critical Study of the Origin, Meanings, and Consequences of the Quantum Principle of Uncertainty*. Unpublished PhD thesis, Institut Supérieur de Philosophie, Université Catholique de Louvin.

Maheu, R. ed. (1971). *Science and Synthesis*. An international colloquium organized by Unesco on the tenth anniversary of the death of Albert Einstein and Teilhard de Chardin. Berlin; Heidelberg; New York: Springer.

Mehra, J. and Rechenberg, H. (1982a). *The Historical Development of Quantum Theory*, vol. 1: The Quantum Theory: Planck, Einstein, Bohr and Sommerfeld: Its Foundation and the Rise of its Difficulties 1900–1925. New York: Springer-Verlag.

Mehra, J. and Rechenberg, H. (1982b). *The Historical Development of Quantum Theory*, vol. 2: The Discovery of Quantum Mechanics 1925. New York: Springer-Verlag.

Mehra, J. and Rechenberg, H. (2000). *The Historical Development of Quantum Theory*, vol. 6: The Completion of Quantum Mechanics 1926–1941: Part I. New York: Springer.

Mehra, J. and Rechenberg, H. (2001). *The Historical Development of Quantum Mechanics*, vol. 6: The Completion of Quantum Mechanics 1926–1941: Part II. New York: Springer.

Miller, A.I. (1978). Visualization lost and regained: the genesis of quantum theory in the period 1913–1927. In *On Aesthetics in Science*, ed. J. Wechsler. Boston: Birkhäuser, pp. 73–102.

Miller, A.I. (1982). Redefining Anschaulichkeit. In *Physics as Natural Philosophy*, ed. A. Shimony and H. Feshbach. Cambridge MA: MIT Press.

Miller, A.I. (1986). *Imagery in Scientific Thought: Creating Twentieth Century Physics*. Cambridge: MIT Press.

Moore, W.J. (1989). *Schrödinger: Life and Thought*. Cambridge: Cambridge University Press.

Muller, F.A. (1997). The equivalence myth of quantum mechanics. *Studies in History and Philosophy of Modern Physics*, **28**, 219–47.

Murdoch, D. (1987). *Niels Bohr's Philosophy of Physics*. Cambridge: Cambridge University Press.

Murdoch, D. (1994). The Bohr–Einstein dispute. In *Niels Bohr and Contemporary Philosophy*, ed. J. Faye and H.J. Folse. Dordrecht: Kluwer Academic Publishers, pp. 303–24.

Neumann, J. von. (1932). *Mathematische grundlagen der Quantenmechanik.* (Die Grundlehren der Mathematischen Wissenschaften, Band 38). Berlin: Springer.

Northrop, F.S.C. (1958). Introduction. In *Physics and Philosophy: The Revolution in Modern Science*, ed. W. Heisenberg. London: George Allen and Unwin Ltd, pp. 11–31.

Pais, A. (1982). Max Born's statistical interpretation of quantum mechanics. *Science*, **218**, 1193–8.

Papenfuß, D., Lüst, D. and Schleich, W.P. eds. (2002). *100 years Werner Heisenberg: Works and Impact*. Weinheim: Wiley-VCH.

Partenheimer, M. (1989). *Goethes Tragweite in der Naturwissenschaft: Herman von Helmholtz, Ernst Haeckel, Werner Heisenberg, Carl Friedrich von Weizsäcker.* Berlin: Duncker & Humboldt.

Pauli, W. (1919). Merkurperihelbewegung und Strahluenablenkung.in Weyls Gravitationtheorie. *Deutsche Physicalische Gesellschaft Verhandlungen*, **21**, 742–50.

Pauli, W. (1926). Quantentheorie. *Handbuch der Physik*, ed. H. Geiger and K. Scheel, **23**, pp. 1–278.

Pauli, W. (1927). Über Gasentartung und Paramagnetismus. *Zeitschrift für Physik*, **40**, 81–102.

Pauli, W. (1958). *Theory of Relativity*. New York: Pergamon Press.

Pauli, W. (1980). *General Principles of Quantum Mechanics*, trans. P. Achuthan and K. Venkatesan. Berlin: Springer-Verlag.

Pauli, W. (1979). *Wissenschaftlicher Briefwechsel mit Bohr, Einstein, Heisenberg*, vol. 1: 1919–1929, ed. A. Hermann, K. von Meyenn and V.F. Weisskopf. New York: Springer-Verlag.

Pauli, W. (1985). *Wissenschaftlicher Briefwechsel mit Bohr, Einstein, Heisenberg*, vol. 2: 1930–1939, ed. K. von Meyenn, with the cooperation of A. Hermann and V.F. Weisskopf. Berlin: Springer-Verlag.

Pauli, W. (1994). The philosophical significance of complementarity [1950]. *Writings on Physics and Philosophy*, trans. R. Schlapp, ed. C.P. Enz and K. von Meyenn. Berlin: Springer-Verlag, pp. 35–42.

Peat, D. and Buckley, P. eds. (1996). *Glimpsing Reality: Ideas in Physics and the Link to Biology*. Toronto: University of Toronto Press.

Perovic, S. (2008). Why were matrix mechanics and wave mechanics considered equivalent? *Studies in History and Philosophy of Modern Physics*, **39**, 444–61.

Petersen, A. (1963). The philosophy of Niels Bohr. *Bulletin of Atomic Scientists*, **19**, 8–14.

Petersen, A. (1968). *Quantum Physics and the Philosophical Tradition*. Cambridge: MIT Press.

Przibram, K. ed. (1967). *Letters on Wave Mechanics*, trans. M.J. Klein. New York: Philosophical Library.

Psillos, S. (2001). Is structural realism possible? *Philosophy of Science*, **68**, 13–24.

192   *References*

Rashevsky, R.V. (1926). Einige Bemerkungen zu Heisenbergschen Quantenmechanik. *Zeitschrift fur Physik*, **36**, 153–8.

de Regt, H. (1997). Erwin Schrödinger, *Anschaulichkeit*, and quantum theory. *Studies in the History and Philosophy of Modern Physics*, **28**, 461–81.

de Regt, H. (1999). Pauli versus Heisenberg: A case study of the heuristic role of philosophy. *Foundations of Science*, **4**, 405–26.

Reichenbach, H. (1920). *Relativitätstheorie und Erkenntnis Apriori*. Berlin: Springer, 1920. Trans. in Reichenbach, 1965.

Reichenbach, H. (1944). *The Philosophical Foundations of Quantum Mechanics*. Berkeley: University of California Press.

Reichenbach, H. (1991). The space problem in the new quantum mechanics [1926]. *Erkenntnis*, **35**, 29–47.

Reichenbach, H. (1996). The present state of the discussion of relativity: a critical investigation. In *Logical Empiricism at its Peak: Schlick, Carnap and Neurath*, vol. 2: Science and Philosophy in the Twentieth Century. Basic Works of Logical Empiricism, ed. S. Sarkar. New York: Garland Publishing, pp. 273–95.

Rosenfeld, L. (1979a). Strife about complementarity. In *Selected Papers of Léon Rosenfeld*, ed. R.S. Cohen and J.J. Stachel. Dordrecht; Boston: D. Reidel, pp. 465–583.

Rosenfeld, L. (1979b). The wave–particle dilemma. In *Selected papers of Léon Rosenfeld*, ed. R.S. Cohen and J.S. Stachel. Dordrecht: D. Reidel, pp. 688–703.

Röseberg, U. (1995). Did they just misunderstand each other? Logical empiricists and Bohr's complementary argument. In *Physics, Philosophy and the Scientific Community*, ed. K. Gavrough, J. Stachel and M.W. Wartofsky. Dordrecht: Kluwer Academic Publishers, pp. 105–23.

Roychoudhuri, C. (1978). Heisenberg's microscope – a misleading illustration. *Foundations of Physics*, **8**, 845–9.

di Salle, R. (1993). Helmholtz's empiricist philosophy of mathematics: between laws of perception and laws of nature. In *Hermann von Helmholtz and the Foundations of Nineteenth-Century Science*, ed. D. Cahan. Berkeley: University of California Press, pp. 498–521.

Sallee, J.C. (1983). Plato and Heisenberg: the humanism of atomic physics. *Proteus*, **1**, 16–22.

Scheibe, E. (1973). *The Logical Analysis of Quantum Mechanics*. Oxford: Pergamon Press.

Schiemann, G. (2008). *Werner Heisenberg*. München: Verlag C.H. Beck.

Schlick, M. (1917). *Raum und Zeit in der gegenwärtigen Physik. Zur Einführung in das Verständnis der allgemeinen Relativitätstheorie*. Berlin: Julius Springer.

Schlick, M. (1979a). Critical or empirical meaning of the new physics. *Philosophical Papers*, vol. 1, ed. H.L. Mulder and B.F.B. van de Velde-Schlick, trans. P. Heath. Dordrecht: D. Reidel, pp. 322–34.

Schlick, M. (1979b). Causality in contemporary physics. In *Philosophical Papers*, vol. 2: 1925–1936, ed. H.L. Mulder and B.F.B. van de Velde-Schlick, trans. P. Heath. Dordrecht: D. Reidel, pp. 176–209.

Schlick, M. (1979c). Quantum theory and the knowability of nature. In *Philosophical Papers*, vol. 2: 1925–1936, ed. H.L. Mulder and B.F.B. van de Velde-Schlick, trans. P. Heath. Dordrecht: D. Reidel, pp. 482–90.

Schlick, M. (1996). Positivism and realism. In *Logical Empiricism at its Peak: Schlick, Carnap and Neurath*, vol. 2: Science and Philosophy in the Twentieth Century. Basic Works of Logical Empiricism, ed. S. Sarkar. New York: Garland, pp. 33–58.

Schröder, W. (1999). *Naturwissenschaft und Religion: Versuch einer Verhältnisbestimmung, dargestellt am Beispiel von Max Planck und Werner Heisenberg*. Bremen: Science Edition.

Schrödinger, E. (1926). An undulatory theory of the mechanics of atoms and molecules. *Physical Review*, **28**, 1049–70.

Schrödinger, E. (1928a). The continuous transition from micro- to macro-mechanics. In *Collected Papers on Wave Mechanics*. London: Blackie and Son Ltd, pp. 41–4. Trans. as J.F. Schearer and W.M. Deans.

Schrödinger, E. (1928b). On the relation between the quantum mechanics of Heisenberg, Born, and Jordan, and that of Schrödinger. In *Collected Papers on Wave Mechanics*. London: Blackie and Son Ltd, pp. 45–61. Trans. as J.F. Schearer and W.M. Deans.

Schrödinger, E. (1928c). Quantisation as a problem of proper values (Part I). In *Collected Papers on Wave Mechanics*. London: Blackie and Son Ltd, pp. 1–12. Trans. as J.F. Schearer and W.M. Deans.

Schrödinger, E. (1928d). Quantisation as a problem of proper values (Part II). In *Collected Papers on Wave Mechanics*. London: Blackie and Son Ltd, pp. 13–40. Trans. as J.F. Schearer and W.M. Deans.

Schrödinger, E. (1930). Zum Heisenbergschen Unschärfeprinzip. *Berliner Berichte*, **19**, 296–303.

Schrödinger, E. (1934). Über die Unanwendbarkeit der Geometrie im Kleinen. *Naturwissenschaften*, **31**, 518–20.

Schrödinger, E. (1957). *Science, Theory and Man*. New York: Dover.

Schweber, S.S. (1994). *Quantum Electrodynamics and the Men who Made It: Dyson, Feynman, Swinger and Tomonga*. Princeton: Princeton University Press.

Serwer, D. (1977). *Unmechanischer Zwang*: Pauli, Heisenberg and the rejection of the mechanical atom 1923–1925. *Historical Studies in the Physical Sciences*, **8**, 189–256.

Shimony, A. (1983). Reflections on the philosophy of Bohr, Heisenberg and Schrödinger. In *Physics, Philosophy and Psychoanalysis*, ed. R.S. Cohen and L. Laudan. Dordrecht: D. Riedel, pp. 209–21.

Shimony, A. (1997). Wigner on foundations of quantum mechanics. In *The Collected Works of Eugene Paul Wigner*. Part A – Vol III: Particles and Fields; Foundations of Quantum Mechanics. Berlin: Springer, pp. 401–14.

Sommerfeld, A. (1927). Zum gegenwärtigen Stande der Atomphysik. *Physicalische Zeitschrift*, **28**, 231–9.

Sommerfeld, A. (1930). Über Unanschaulichkeit in der modernen Physik. *Unterrichtsblätter für Mathematik und Naturwissenschaft*, **36**, 161–7.

Squires, E. (1994). *The Mystery of the Quantum World*. Bristol: Institute of Physics Publishing.

Stapp, H. (1972). The Copenhagen interpretation. *American Journal of Physics*, **40**, 1098–116.

Stepansky, B.K. (1997). Ambiguity: aspects of the wave–particle duality. *British Journal for the History of Science*, **30**, 375–85.

Tanona, S. (2004). Uncertainty in Bohr's response to the Heisenberg microscope. *Studies in the History and Philosophy of Modern Physics*, **35**, 483–507.

Taylor, C. (1985). Theories of meaning. In *Human Agency and Language*. vol. 1: Philosophical Papers. Cambridge: Cambridge University Press, pp. 248–92.

Thiring, H. (1928). Die Grundgedanken der neuen Qauntentheorie, erster Teil: Die Entwicklung bis 1926. *Ergebnisse Exakten Naturwissenschaften*, **7**, 384–431.

Uffink, J. and Hilgevoord, J. (1985). Uncertainty principle and uncertainty relations. *Foundations of Physics*, **15**, 925–44.

Waerden, B.L. van der. ed. (1967). *Sources of Quantum Mechanics*. Amsterdam: North-Holland.

Waerden, B.L. van der. (1973). From matrix mechanics and wave mechanics to unified quantum mechanics. In *The Physicist's Conception of Nature*, ed. J. Mehra. Dordrecht: D. Reidel, pp. 276–93.

Weizsäcker, C.F. von. (1931). Ortsbestimmung eines Elektrons durch ein Mikroskop. *Zeitshrift für Physik*, **70**, 114–30.

Weizsäcker, C.F. von. (1952). *The World View of Physics*, trans. M. Grene. London: Routledge.

Weizsäcker, C.F. von. (1955). Komplementarität und Logik. *Die Naturwissenschaften*, **42**, 521–9, 545–55.

Weizsäcker, C.F. von. (1963). *Zum Weltbild der Physik*. Stuutgart: Hirzel.

Weizsäcker, C.F. von. (1971a). The Copenhagen interpretation. In *Quantum Theory and Beyond*, ed. T. Bastin. Cambridge: Cambridge University Press, pp. 25–31.

Weizsäcker, C.F. von. (1971b). Notizen über die philosophische Bedeutung der Heisenbergschen Physik. In *Qaunten und Felder, physicalische und philosophische Betrachtungen zum 70. Geburtstag von Werner Heisenberg*, ed. H.P. Dürr. Branschweig: F. Vieweg, pp. 11–26.

Weizsäcker, C.F. von. (1985). A reminiscence from 1932. In *Niels Bohr: A Centenary Volume*, ed. A.P. French and P.J. Kennedy. Cambridge: Harvard University Press, pp. 183–90.

Weizsäcker, C.F. von. (1987). Heisenberg's philosophy. In *Symposium on the Foundations of Modern Physics*. The Copenhagen Interpretation 60 years after the Como Lecture, ed. P. Lahti and P. Mittelstaedt. Singapore: World Scientific, pp. 277–93.

Weizsäcker, C.F. von. (1988). *The Ambivalence of Progress: Essays on Historical Anthropology*. New York: Paragon House.

Weizsäcker, C.F. von. (1994). Kant's theory of natural science according to P. Plaass. In *Kant's Theory of Natural Science*, ed. P. Plaass, with introduction and commentary by A.E. Miller and M.G. Miller. Dordrecht: Kluwer, pp. 167–87.

Wessels, L. (1980). What was Born's statistical interpretation of quantum mechanics? In *Philosophy of Science Association, Proceedings of the 1980 Biennial Meetings of the Philosophy of Science Association*, vol. 2, ed. P.D. Asquith and R.N. Giere. East Lansing, MI: Philosophy of Science Association, pp. 187–200.

Wessels, L. (1993). Erwin Schrödinger and the Descriptive Tradition. In *Springs of Scientific Creativity. Essays on the Founders of Modern Science*, ed. R. Aris, H.T. Davis and R.H. Stuewer. Minneapolis: University of Minnesota Press, pp. 254–78.

Weyl, H. (1920). Elektrizität und Gravitation. *Physicalishe Zeitschrift*, **21**, 649–50.

Weyl, H. (1949). *Philosophy of Mathematics and Natural Science*, based on Trans. O. Helmer. Princeton: Princeton University Press.

Wigner, E.P. (1983a). The problem of measurement. In *Quantum Theory and Measurement*, ed. J.A. Wheeler and W.H. Zurek. Princeton: Princeton University Press, pp. 341–4.

Wigner, E.P. (1983b). Remarks on the mind-body question. In *Quantum Theory and Measurement*, ed. J.A. Wheeler and W.H. Zurek. Princeton: Princeton University Press, pp. 168–81.

Worrall, J. (1989). Structural realism: the best of both worlds? *Dialectica*, **43**, 99–124.

Index

Anschaulichkeit 5, 37, 48–53, 59, 173
  Heisenberg's redefinition of 37, 48–53
  and matrix mechanics 33, 38
  Schrödinger's view of 48
A priori, redefinition of 5, 6, 9, 134, 135,
  137–138, 142–143, 147–151, 173, 174
Aristotle 11, 167–168

Bauer, E. 168–169
Beller, M. 2, 4, 7, 25, 30, 49, 52, 54, 64, 85, 103,
  117, 151, 176, 177
Bohm, D. 65, 159
Bohr, N. 1, 65, 76, 134, 136, 142, 166, 172, 174,
  175, 176, 178
  and Kantian philosophy 2, 133, 141, 143
  and positivism 2, 155
  and subjective idealism 2
  complementarity. *See* complementarity
  discussions with Schrödinger. *See*
    Schrödinger
  inter-phenomenal object 165
  intuitive understanding 48, 52–53, 58
  object-instrument entanglement. *See* object-
    instrument divide
  on language 154–155
  quantum theory of atomic structure 22–24,
    27, 28, 36
  realism and anti-realism 8
  unambiguous description 120–122, 128, 177
  wave-particle duality. *See* wave-particle
    duality
Bokulich, A. 9, 55, 56
Born, M. 17, 19, 21, 22, 33, 38, 39, 44, 52, 55,
  65, 77, 81, 87, 89, 117, 119, 123
  on observability 17, 20–21, 29–30

  on statistical interpretation of quantum
    mechanics 43, 46, 68–69, 72, 74, 114
Bridgman, P. 25, 92
de Broglie, L. 65, 67, 74, 75, 77, 175
  wave theory of matter. *See* matter waves
de Broglie waves 68, 74, 75
Bromberg, J. 73
Buckley, P. 153, 154, 156, 164

Carella, M. 167
Carnap, R. 57, 134, 136, 137–138, 150, 155
Carson, C. 11
Cassidy, D. 4, 10, 22, 26, 32, 142
Cassimir, H. 163, 165
Cassirer, E. 12, 19, 51, 57, 88, 106, 134, 138,
  143, 152, 157–158, 160–161
Causality in quantum mechanics 54, 110, 113–115
Chevalley, C. 2, 4, 9, 12, 13, 50, 147, 158, 172,
  175, 177
Classical concepts. *See* Doctrine of
  indispensability of classical concepts
Closed theories. *See* Heisenberg
Coffa, A. 137
Complementarity x, 2, 5, 63, 108–129, 142, 177
  Bohr on space-time and causal descriptions
    108, 109–112
  Heisenberg on space-time and causal
    descriptions 112–118
  kinematic-dynamic complementarity 109,
    118, 119–122, 128
  mutually exclusive experimental
    arrangements 108, 112, 118–121
  stationary states 108, 111–112
Completeness of quantum mechanics 122–126,
  163, 177–178

Copenhagen interpretation ix, x, 1–2, 104, 110, 128, 140, 159, 166, 175, 177
Crull, E. 125
Cushing, J. 46

Darrigol, O. 25, 73
Darwin, C. 71
Davisson, C. 80
Degen, A. 175
Descartes, R. 11
Dirac, P. 2, 17, 39, 40, 41, 65, 67, 83, 97, 114
  On second quantisation 73, 74
Dirac-Jordan transformation theory 46, 69, 70, 85
di Salle, R. 50
Doctrine of indispensability of classical concepts 7, 86, 101–102, 133, 140–141, 145–146, 148–150, 151, 153, 174
Drude, B. 30–31
Duns Scotus 11

Eddington, A. 92
Ehrenfest, P. 71
Einstein, A. 1, 4, 8, 21, 29, 31, 43, 44–45, 58, 67, 71, 81, 159, 163, 172, 175
  debate with Bohr over quantum mechanics 120–121, 123
  on intuitive understanding 51–52
  theory of relativity x, 5, 17, 19–20, 35, 36, 48, 92–95, 100, 101, 103–104, 135, 138, 173
Einstein-Podolsky-Rosen paper 119–120

Faye, J. 165
Fichte, J. 11
Fock, V. 155
Folse, H. 141, 165
Forman, P. 49
Fowler, P. 66
Frank, P. 52, 143, 155
Frappier, M. 9, 55, 85, 105, 122
Frege, G. 155, 161
Friedman, M. 137, 150

Gadamer, H. 157
Gamma-ray microscope 6, 86–87, 92–102, 104–105
Gembillo, G. 8, 10, 13
Germer, H. 80
Gieser, S. 1
Goethe, W. 6
Gower, B. 57

Hamann, J. 12, 152, 158
Heelan, P. 8, 9, 11, 13, 19, 113, 127, 134, 136, 176
Hegel, G. 11
Heidegger, M. 11–12, 157, 159, 173
Heisenberg, W.
  1942 philosophical manuscript 12–13, 144, 148, 152, 162, 167, 172
  and German philosophy 4, 145, 152, 157–158, 171, 172
  and Heidegger 11–12
  and Kantian philosophy ix, 5, 6, 9, 129, 133–151, 173
  and neo-Platonism ix, 10, 58
  and positivism x, 5, 17–18, 34, 53–58, 63, 106, 173, 174, 175
  and realism 6, 37, 53–58, 176
  closed theory 6, 55–57, 122
  complementarity. *See* complementarity
  completeness of quantum mechanics. *See* completeness of quantum mechanics
  concept of objective reality x, 7, 159–161, 163–164, 166, 170–171, 174–175
  criticism of wave mechanics 40–43, 66–68, 74
  debate with Schrödinger 5, 36, 49, 58
  discussions with Bohr x, 4, 5, 6, 39, 63, 87, 176
  discussions with Einstein 6, 32–33
  discussions with Weizsäcker. *See* Weizsäcker
  indeterminacy principle. *See* uncertainty relations
  on elimination of electron orbits 23–25, 173
  on intuitive understanding 36, 45–48, 58
  on observability. *See* observability principle
  on the 'cut'. *See* object-instrument divide
  on visualisability. *See* *Anschaulichkeit*
  philosophy of language 6–7, 144–147, 152–171, 172, 173, 174, 175, 178
  Physics and Philosophy 11, 140, 144, 163, 167
  potentiality or *potentia* x, 7, 134, 151, 153, 165–171, 175
  probability interpretation of wave mechanics 70–72
  theory of meaning 91–95, 161–164, 173
  unified field theory 3, 10, 58
  wave-particle duality. *See* wave-particle duality
  wave-particle equivalence 7, 59, 64, 75–77, 83–84
Heitler, W. 135
Helmholtz, H. 12, 152, 158
Hendry, J. 2, 4, 53, 99, 177
Hentschel, K. 24

Hermann, G. 6, 9, 133, 142, 151, 172, 174
Herz, H. 24
Hilgevoord, J. 85
Hoag, B. 55
Honner, J. 121
Hörz, H. 8, 10, 13
Howard, D. ix, x, 2, 7, 8, 108, 127–128, 137,
    148, 150, 167, 177
Humboldt, W. 12, 152, 157–158
Hund, F. 72
Husserl, E. 9, 12, 134, 137

Indeterminacy principle. *See* Uncertainty
    relations
Inter-phenomenal object in quantum mechanics
    165–167

Jammer, M. 2, 28, 98, 103, 110
Jordan, P. 65, 77, 87, 91, 155, 175
    complementarity 108, 109
    matrix mechanics 38, 39, 42, 43, 44–45
    observability 17, 19, 21
    quantum electrodynamics 72, 73–76, 82–83

Kaiser, D. 141
Kant, I. ix, 5, 6, 11, 121, 133–138, 141–144,
    145–147, 157
Kantian philosophy 2, 4, 9, 44, 50–52, 59, 158
Kennard, E. 70, 71
Klauss, F. 13
Klein, O. 68, 72, 74, 75–77, 78, 82–83, 110
Kojevnikov, A. 90
Kramers, H. 52, 66, 111
Kronig, R. 25, 66
Kuby, E. 58
Kuhn, T. 13, 21, 30, 32, 34, 41, 45, 75, 76, 79, 80,
    82, 84, 97, 101–102, 104, 117, 149, 167

Ladyman, J. 57
Lafont, C. 164
Language
    analytic *versus* transcendental conceptions 6,
        135, 151, 152
    designative, or referential, function of 152,
        161–164
    world-disclosing, or constitutive, function of
        152, 154, 161, 171, 173
von Laue, M. 72, 88
London, F. 168–169
Leibniz, G. 155
Lewis, C. 134

Lorentz, H. 43, 65, 70
Lorenz, K. 150

Mach, E. 11, 19, 29, 137
MacKinnon, E. 25, 38, 99, 156, 178
Matrix mechanics 21, 27, 36
    Heisenberg's interpretation of 37–40, 46–47
Matter waves
    de Broglie 68, 74, 75
    quantised matter waves 75–77
    Schrödinger 66
Maxwell's theory, 75–77
Mehra, J. 19, 73
Miller, A. 50
Muller, F. 41
Murdoch, D. 1, 82, 120

Nelson, L. 142
Neo-Kantian philosophy. *See* Kantian
    philosophy
von Neumann, J. 2, 168–169
Northrop, F. 114

Object-instrument divide in quantum
    mechanics 123–128, 133, 139–140
Observability principle 5, 17–35, 173
    as guiding principle 19, 22, 25–26, 27
    hitherto unobservable *versus* unobservable in
        principle 27–31
    observability and non-observability 33–34
Operationalism 6, 23, 33, 38, 63, 86, 92–95,
    101, 103, 153, 173–174
Ordnung der Wirklichkeit. See Heisenberg 1942
    philosophical manuscript
Orseen, 119

Pais, A. 43
Pauli, W. 4, 8, 19, 22, 25, 39, 41, 42, 43, 44–45,
    47, 52, 65, 75, 77, 78, 86, 172
    complementarity 108, 109, 118, 119
    correspondence with Heisenberg on
        uncertainty relations 88–94, 96, 97, 103
    on elimination of electron orbits 22–23
    on observability 17, 18, 19–20, 29, 92
Peat, D. 103, 153, 154, 156
Perovic, S. 41
Petersen, A. 31, 154–155
Planck, M. 1, 10, 72
Plato 11, 58
Positivism x, 137, 142, 144, 175
Psillos, S. 57

Quantum concepts of space-time, 87–90,
100, 174
Quantum electrodynamics 3, 59, 72, 73–76
Quine, W. 134

Rashevsky, R. 89
Rechenberg, H. 19, 73
de Regt, H. 23, 48
Reichenbach, H. 12, 38, 51, 90, 134, 136, 137,
150, 175
Rosenfeld, L. 82, 155
Rupp, E. 80
Russell, B. 12, 57, 155–156

Schiemann, G. 10
Schlick, M. 11, 51, 57, 106, 119, 134, 136, 137,
150, 155–156, 172, 173
Schrödinger, E. 1, 4, 5, 8, 34, 35, 36, 40, 71, 72,
73, 74, 88, 110, 123, 148, 159, 172, 175
discussions with Bohr in Copenhagen 45,
63, 66
on understanding in physics 43, 44–45, 46
on visualisability. *See Anschaulichkeit*
wave mechanics 40–43, 65, 66
Serwer, D. 27
Shimony, A. 141, 167–168
Slater, J. 111
Soler, L. 142
Sommerfeld, 49
Squires 175
Stapp, H. 156
Stepansky, B. 63
Strauss, M. 12
Structural realism 6, 37, 55, 56–58, 176
Subjectivist interpretation of quantum
mechanics 2, 135, 168–170

Taylor, C. 161
Thirring, H. 17
Thomas Aquinas 11
Thomson, G. 80

Uffink, J. 85
Uncertainty relations. 5, 63, 70, 85–106,
114, 120
Bohr's analysis of measurement 95–96,
106, 139
discussions between Bohr and Heisenberg
96–100, 101–102, 177

van der Waerden, B. 39
Visualisability. *See Anschaulichkeit*

Wave packets, and probability packets 70–71, 112
Wave-particle duality x, 5, 59, 63–84, 142, 177
disagreement between Bohr and Heisenberg
96–100, 101–102
Weizsäcker, C. F. 4, 6, 9, 10, 105, 127, 134, 136,
166, 172
discussions with Heisenberg in Leipzig 133,
134, 142, 151, 174, 178
Kantian philosophy 143–144, 149–150
complementarity 116–117
Wessels, L. 69
Weyl, H. 8, 31, 51, 92, 134, 138
Wien, W. 44, 45
Wigner, E. 2, 72, 74, 75–77, 82–83, 135, 169
William of Ockham 11
Wittgenstein, L. 11, 12, 134, 155–156
Wheeler, J. 46
Worrall, J. 57

Zurek, W. 46

Printed in the United States
By Bookmasters